Hares

Nancy Jennings

BLOOMSBURY
LONDON • OXFORD • NEW YORK • NEW DELHI • SYDNEY

Bloomsbury Publishing

50 Bedford Square
London
WC1B 3DP
UK

1385 Broadway
New York
NY 10018
USA

www.bloomsbury.com

First published 2017

British Library Cataloguing-in-Publication Data
A catalogue record for this book is available from the British Library.

Library of Congress Cataloguing-in-Publication data has been applied for.

ISBN: PB: 978-1-4729-3364-5
ePDF: 978-1-4729-3366-9
ePub: 978-1-4729-3365-2

2 4 6 8 10 9 7 5 3 1

Design by Susan McIntyre
Maps by Julie Dando (p10 and p15) and Brian Southern (p22 and p33)
RR Donnelley Asia Printng Solutions Limited

Contents

Meet the Hares

Hares, though not often seen, are characteristic animals that are known and loved by many. There is something magical about hares – they feature in folklore, art, children's books and poetry. The hare is a symbol of fertility, resurrection and immortality, perhaps because it has the ability to pop up unexpectedly as though suddenly born or reborn, appearing out of nowhere, only to disappear from view by sprinting away.

Only a handful of our mammal species are familiar to most of us, and not many people are lucky enough to see wild mammals regularly. Depending on where you live, you might see Foxes, squirrels, rats, mice, deer, bats and Rabbits. Though they are more rarely seen, hares are among our most distinctive mammals. Some might describe them as ungainly, long-legged rabbit-like animals with oversized ears. Hares are similar to Rabbits, but they are bigger and faster, and also more timid, elusive and elegant than the much more familiar Rabbits. The hare is an unusual species in that it breeds quickly and does not live long, but is relatively large (see Speedy lifestyles, page 45). It is also a game animal and food source that is eaten by humans and other animals (see Eat or Be Eaten, page 77). In some parts of the world, hares are invasive pests, causing harm to fragile ecosystems (see Conservation of Hares, page 89). In others, the conservation of hares presents a paradox: why are hares threatened by intensive farming, though they thrive in arable areas?

Opposite: Brown Hares have oversized ears, large eyes, and long, sensitive whiskers: adaptations for avoiding predators.

Below: Arable fields provide food and shelter for hares, while field margins allow them to add diversity to their diet.

Above: The 'common hare', or Brown Hare, as depicted by the artist A. Thorburn in *British Mammals*, published in 1920.

Below: A Mountain Hare in its winter coat is well camouflaged in snow. Illustration by A. Thorburn, *British Mammals*, 1920.

Two (or more) British species

Two species of hare are widespread in Europe: the Brown (or European) Hare and the Mountain Hare. There are 15 European subspecies of Mountain Hare. The Mountain Hares in the British Isles belong to two subspecies: one is found in Scotland and the Peak District; the other is found in Ireland, is called the Irish Hare and may in fact be a separate species. There are also at least 16 subspecies of Brown Hare, which differ in colour, size, and skull and tooth shape, but their relationships and distributions are still being clarified by scientists. This book focusses on the Brown Hare, the Scottish Mountain Hare and the Irish Hare. These species are similar enough to be difficult to identify when seen in isolation but, as they are found in different areas and habitats, you are unlikely to confuse them. If you were to see two of these hares side by side – which could happen only with a great deal of luck or skill, and only in a handful of locations – you could tell them apart easily enough with careful observation.

Below left: The Brown Hare is distinctly reddish-brown. In summer, the Mountain Hare (**right**) is similar in colour, but its winter coat may be partly or entirely white. This individual, photographed in spring, is moulting to become brown again.

The Brown Hare

Above: The long back legs of Brown Hares allow them to move swiftly.

The Brown Hare (*Lepus europaeus*), though fairly common on farmland in Europe, is tricky to see because it is secretive, elusive and nocturnal or crepuscular. Adult Brown Hares in England weigh on average 3.5kg (7¾lb), as much as a domestic cat, and are about 55cm (22in) long from nose to tail. Their back feet are very large, at about 15cm (6in) long. The back legs are much longer and larger than their front legs; as a result, hares have a peculiar galloping gait, in which the front feet land first and the hind feet pass outside the front feet and land in front of them. The ears appear to be oversized at about 10cm (4in) long. If you fold the ears of a Brown Hare forward, over the top of its nose, the ears stick out a little beyond the tip of the nose. This is not the case with a European Rabbit, or even with an Irish or Mountain Hare. What immediately springs to mind, of course, is the astonishing fact that Bugs Bunny is actually a hare!

Brown Hares live their whole lives above ground, creating forms or seats (depressions in the ground or vegetation) to shelter in during the day. They rely on speed to escape from their main natural enemy, the Red Fox, and from other predators, as hares do not use burrows. Brown Hares are mainly solitary, though they come together to

Below: Bugs Bunny's long ears and limbs make him look more like a hare or a jackrabbit than a rabbit.

mate and sometimes to feed. During the mating season, several potential suitors (males, also called bucks) follow a female (or doe) around, and females who are not ready to mate 'box' the males, simultaneously pushing them away and testing their strength.

Though Brown Hares may look patchy and scruffy when they moult (in spring and autumn), their coats otherwise change little over their lives. They are born with fur very close in colour to the adult coat and their summer and winter coats look similar. When you glimpse a Brown Hare, especially in sunlight, the most striking identification feature is its colour. It is distinctly reddish brown on the back, though the red is more patchy and more brindled or grizzled than the red of a Fox, because each individual hair is banded in different colours along its length. The coat becomes yellower on the flanks, on the sides of the face and on the inner limbs. The belly is creamy white; the tail white with black on top. Wild European Rabbits, on the other hand, appear much more even in colour, and are more greyish brown than reddish brown. There are grey, black, white and sandy-coloured forms of the Brown Hare, but they are very rare.

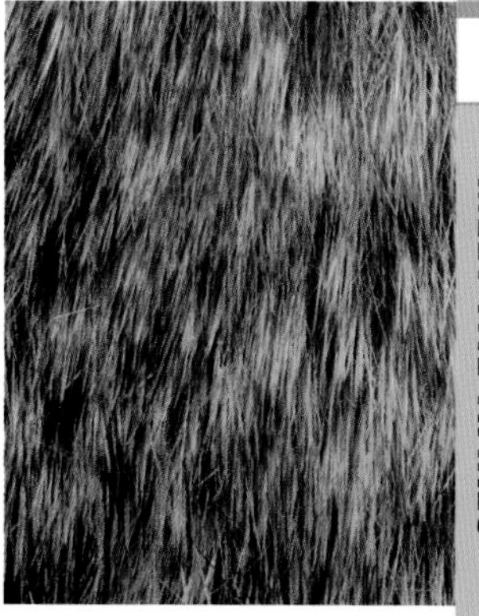

Above: Each hair in the fur of a Brown Hare has bands of several colours, giving the animal a brindled appearance.

In sunlight, Brown Hares usually settle in their forms, where they are hard to spot as they are well camouflaged and motionless. Try scanning a field with binoculars, looking for something that resembles a lump of manure or mud – it might turn out to be a hare. Once you spot a hare in its form, you may be able to approach quite closely and watch it sink gradually lower and lower into the ground, until you get so close that the hare finally decides to run for it. Then you will see the back end of the hare disappearing very quickly and, if you can find it, you may see the still warm and cosy-looking form. Usually the form provides the hare with a good view over

Left: A Brown Hare keeping watch from its form. From a distance, it can resemble a clump of mud.

any open areas. If you watch Brown Hares when they are most likely to be active and visible, at dawn or dusk and in low light, the pale margins of their downturned tails and the back of their ears stand out.

Range and habitats

The Brown Hare is found throughout the UK, including on several surrounding islands. It has a wide native (natural) range in Europe and Asia, stretching as far as Lake Baikal in Siberia. Over the years, Brown Hares have been caught (or bred in captivity), boxed up, moved around and released into the wild many times, mainly to create new populations or to increase existing populations for hunting. Introduced populations of Brown Hares now live outside their native range, in Northern Ireland, Siberia, Finland, Sweden, eastern Canada, north-eastern USA, southern South America, Australia, New Zealand, Barbados, Réunion (near Madagascar), the Bahamas and the Falkland Islands. The Brown Hare is also an invasive non-native species in many places.

In Australia, there were 30 known attempts to introduce Brown Hares, of which 14 were successful. Brown Hares became established in Tasmania in 1837

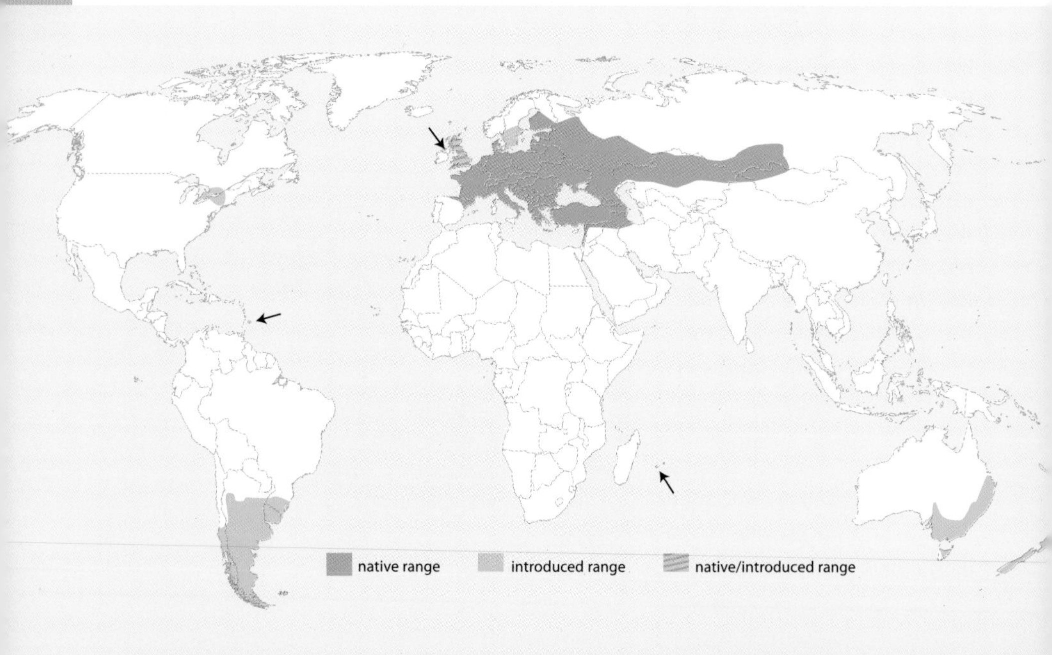

Below: Brown Hare world distribution map, showing the native and introduced range, and the British Isles, where it is unknown whether the Brown Hare is native or introduced. The arrows indicate small introduced populations in Ireland, Barbados and Réunion.

and on the mainland in 1855. They spread naturally at a rate of less than 2km (1¼ miles) per year, and in 1867 the Acclimatisation Society of Victoria began distributing hares to the landed gentry throughout Victoria and New South Wales. The last known introduction of Brown Hares to Australia was in 1872. By 1900, they had reached the Queensland border and had become agricultural pests in Victoria and New South Wales.

Though the Brown Hare is classed as Near Threatened or Threatened in several individual countries, it is classed as Least Concern in the Red List of the International Union for Conservation of Nature (IUCN). This indicates that globally, it is not considered threatened, vulnerable or endangered. Like other hare species, Brown Hares rely on being able to see predators, and avoid them by running fast, so they prefer flat open habitats with short vegetation, especially at night, when they are feeding. They can be found in saltmarshes, arid areas, moorland, alpine grassland, steppe (semi-arid grassland), pampas (South American fertile plains) and sand dunes. However, in Europe they are most common in arable areas used for cereal growing.

Above: Brown Hares in Europe generally like arable crops, but need other habitats for shelter and food at different times of year.

Brown Hares do need some permanent cover, and leverets (young hares) survive better in mixed agricultural areas than in cereal monocultures. Therefore, large-scale intensive farming, with a low diversity of crops and large fields, is less suitable for them than farming that provides food and shelter all year round in small fields. Brown Hares like strips of uncultivated land in arable fields; creating such strips and windbreaks, and increasing crop diversity, can lead to dramatic increases in numbers of hares. Woods, shelterbelts and hedgerows are used by Brown Hares for resting and occasional feeding during the day, particularly in winter. They will feed in pasture all year round, but may do so more in summer when cereal crops are no longer edible for hares. High densities of cattle and sheep may deter Brown Hares.

The Brown Hare can tolerate mean annual temperatures of 3–30°C (37–86°F) and annual rainfall of 200–1,200mm (7¾–47in), and occupies elevations of up to 2,500m (8,200ft) above sea level.

Introduced species and acclimatisation societies

Acclimatisation societies existed mainly during the colonial era; their aim was to move species around the world. The first was formed in Paris in 1854. In 1860, an acclimatisation society was formed in London and a year later Peafowl, Pheasants, swans, Starlings and Linnets had been introduced to Australia. The Acclimatisation Society of Victoria, Australia, was formed in 1861.

The Australian colonies were considered to have 'impoverished' faunas that could be improved by introducing species to recreate the fauna and flora of the colonists' homelands. Skylarks, Goldfinches, Pheasants, House Sparrows, thrushes, Mallards, Brown Trout, Cane Toads, Cashmere Goats, Alpacas, deer, Brown Hares and many other species were introduced to Australia. Foxes were introduced for hunting in the 1830s and 1840s; there are now over six million of them, and they have caused the extinction of several native Australian species. European Rabbits were introduced to Australia as food and became widespread in the 1850s. They were successfully introduced in New Zealand in the 1830s, but by 1876 the Rabbit Nuisance Act had been passed, showing that the devastation they caused had been understood. They munched away and multiplied, causing the loss of many plant species, and soil washed away in areas that had been overgrazed by Rabbits. Stoats, later introduced to control Rabbits and Brown Hares, had little effect on the target species and are now a major threat to the native birds of New Zealand!

To understand how commonly humans have moved species, just think about how many familiar animals are actually not native to the British Isles: Grey Squirrels are from the eastern USA, Rabbits are from Spain and Portugal, Fallow Deer are from southern Europe, Little Owls are from mainland Europe, and Pheasants are from Asia.

We now understand the devastation that can be caused to ecosystems by the introduction of non-native plant and animal species. Such species are often invasive (harmful in their new environment): they may carry diseases, outcompete native species, or prey upon them. There are almost 2,000 invasive non-native species in Great Britain. Acclimatisation turned out to be a big mistake, and thankfully societies devoted to the introduction of species to new areas have had their day. In fact, already about 20 years after the peak popularity of acclimatisation, most of the societies had folded, and those remaining were managing zoos and botanical gardens.

The Mountain Hares

Left: A Mountain Hare in its winter coat, almost invisible against the snow.

The Mountain Hare (*Lepus timidus*) has a body shape similar to the Brown Hare. The two species are so closely related that they have been assigned to the same genus, *Lepus*. The Mountain Hare, too, has longer hind legs than front legs, a fast galloping gait, long ears and does not burrow, though occasionally leverets make and use shallow burrows. It is mostly solitary, avoids predators by moving fast, and is crepuscular or nocturnal. It is tougher than the Brown Hare – adapted for northerly, mountainous and polar areas – and has slightly longer back feet with thick fur and spreading toes for moving on snow. Its face is more convex than the Brown Hare's, its ears lack a white patch, and its tail is all white.

The Mountain Hare moults two or three times each year (three times in Scotland: February–May, white to brown; June–September, brown to brown; October–February, brown to white). Its summer coat is brown with blue-grey underfur, but in winter it may remain brown, turn white or partly white, or turn blue-grey. Understandably it is also called the Blue Hare, Tundra Hare, Variable or Varying Hare, White Hare, Snow Hare or Alpine Hare. Black,

Above: Irish hares on a rainy golf course in Straffan, County Kildare, Ireland. Most Irish Hares remain brown all year round.

white and sandy-coloured forms of the Mountain Hare are possible but very rare.

The Mountain Hare in Scotland (subspecies *Lepus timidus scoticus*) is similar to the 13 subspecies found in the rest of Europe: nearly all individuals turn white in winter. The Irish Hare (subspecies *Lepus timidus hibernicus*) is more distinct from other European subspecies and may prove to be a separate species. Irish Hares often remain reddish brown all year, though some individuals turn partly white in winter (white fur is usually on the rump, flanks and legs; the back and head remain brown).

Range and habitats

The Mountain Hare is found throughout Eurasia, from Norway to Japan, with a stronghold in Russia, Scandinavia and the Baltic States. Further around the globe, it is replaced by two closely related species: the Alaskan or Tundra Hare (*Lepus othus*) in Alaska, and the Arctic Hare or Polar Rabbit (*Lepus arcticus*) in Greenland and the Canadian Arctic. Some taxonomists believe that the Mountain Hare, Alaskan Hare and Arctic Hare are the same species.

The Mountain Hares in Ireland, Scotland and the Alps are populations that have survived there since after the last ice age and are now separated from the main population. Other isolated populations exist in Hokkaido (Japan), the

Kurile Islands and Sakhalin (Russia). The populations in southern Scotland, some Scottish islands, the Faroe Islands (Denmark), the Isle of Man and the Peak District (UK) were all introduced by humans, mainly in the 19th century. In the Isle of Man and the Peak District, Brown Hares are more common on farmland and Mountain Hares are more common on heather moorland. Irish Hares were introduced to south-western Scotland, and Scottish Mountain Hares were introduced to Ireland and to the Isle of Man. So, as with the Brown Hare, the locations where Mountain Hares are found today are a result of intervention by humans over many decades, as well as a result of natural processes.

Numbers of Mountain Hares in most areas are stable, and the species is classed as Least Concern by the IUCN (not threatened, vulnerable or endangered). However, numbers have fluctuated in northern Europe and declined in the Alps. Sometimes parasites, predation or starvation cause sudden crashes in the population. Declines have occurred in Russia, and in southern Sweden the Mountain Hare has become locally extinct where the Brown Hare has invaded, habitats have changed, or snow cover has been reduced due to climate change. Competition with the Brown Hare also occurs in Northern Ireland, where

Above: A Scottish Mountain Hare in its favourite habitat, heather moorland.

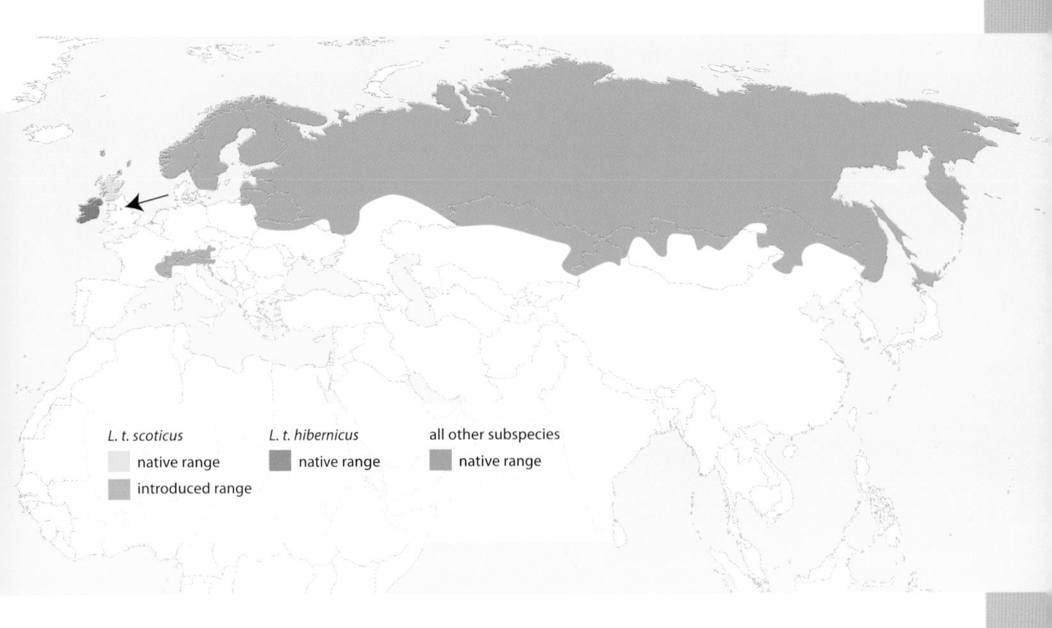

Below: Mountain Hare world distribution map, showing the subspecies found in Ireland and in Scotland separately, and all 13 other subspecies together. The arrow indicates a small introduced population in the Peak District

Above: Mountain Hares use forms to provide shelter during the day.

historical records of the numbers of Irish Hares shot (game bag data) suggest that numbers are declining.

The Mountain Hare mostly inhabits mixed forest (pine, birch and juniper), clear-felled areas, open glades, swamps and bogs, river valleys, moor, tundra (a habitat in which small trees are few and far between, moss and heath grows, and the ground is often permanently frozen), and taiga (northern coniferous forests), up to the elevational limit of vegetation (the 'tree line'). In Scotland heather moorland, forests and pasture are the preferred habitats. In Scotland, Mountain Hares make their forms in heather, snow or peat; even these shallow depressions shelter Mountain Hares effectively from the wind. Some forms are used for many years.

Irish Hares are found in a wide variety of habitats, from the coast where they sometimes eat seaweed, to cereal fields, grassland and the tops of mountains, though they are most common in agricultural grasslands. This means that, in parts of Ireland, Irish Hares and Brown Hares live in the same habitats and compete for food. Mountain Hares feed on many different plants, including grasses, sedges, mosses, trees and shrubs, as well as lichens, but heather is their staple diet in Scotland.

Hares and rabbits

Left: Scottish Mountain Hare in summer, in its favourite habitat, heather moorland.

The species of hare this book focusses on are closely related: they belong to the genus *Lepus* and have similar biology, appearance and habits. The European Rabbit belongs to a different genus in the same family (Leporidae) and is perhaps more familiar. A comparison between the basic biology of the European Rabbit, the Brown Hare and the Mountain Hare (Scottish and Irish) shows that, though all of them breed 'like rabbits', hares have very different social lives – they occupy individual forms and don't socialise as much as Rabbits, which live in complex colonial burrow systems. Hares are built for

Left: European Rabbits live in sociable groups in burrows that are often dug into sandy soil.

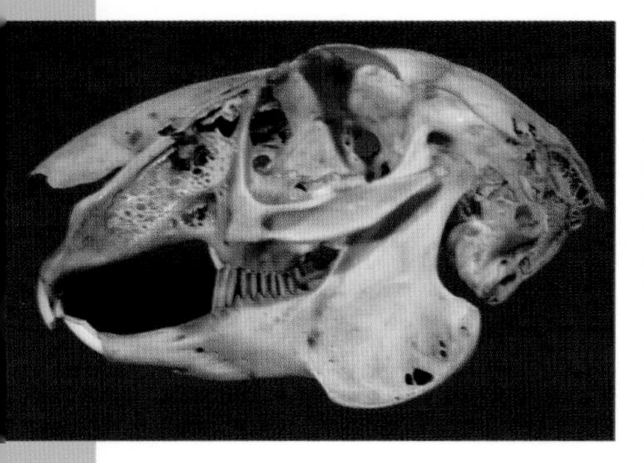

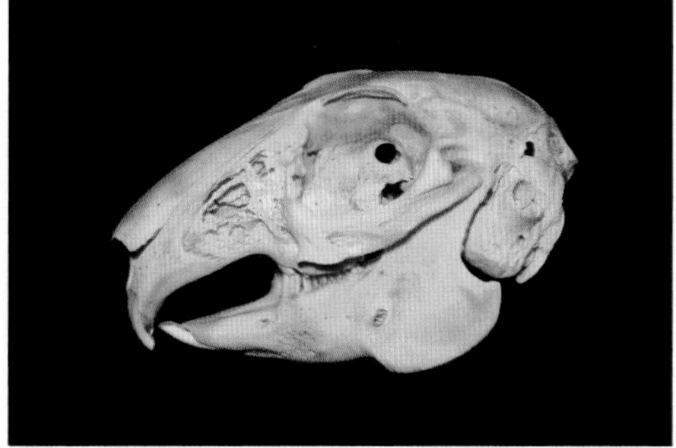

Above: A Brown Hare skull (**left**: around 10cm long) and a European Rabbit skull (**right**; around 7 cm long). The skulls of both species have four incisors in the upper jaw and the bone is delicate in structure so that the skulls are relatively light-weight.

speed, and have evolved longer legs, more blood and relatively bigger hearts than Rabbits. Though hares have big ears and excellent hearing, they keep quiet unless threatened, and don't often use calls to communicate with each other, instead relying on scent and vision. Rabbits use sound more: they thump their feet to raise the alarm.

Baby Rabbits are usually born in short burrows that are not connected to the main burrow system. Their mother builds a nest made of grass or moss and lined with her own fur, and once she has given birth she leaves her young together, but sealed up in the burrow most of the time, only returning to feed them. Baby Rabbits are born naked and with their eyes closed, while leverets are born relatively mature and mobile, and can survive without a burrow.

Below: Rabbits are born blind and naked, but Brown Hares are born fully furred and ready for the world.

Hares and Rabbits compared

	Brown or European Hare	Scottish Mountain Hare	Irish Hare (Mountain Hare)	European Rabbit
Scientific name	*Lepus europaeus*	*Lepus timidus scoticus*	*Lepus timidus hibernicus*	*Oryctolagus cuniculus*
Social life	Solitary	Solitary	Solitary	Social, colonial, live in warrens
Shelter	Forms	Forms in heather and peat	Forms	Burrows
Main food	Herbs, grass, cereals	Heather	Grass	Grass
Average adult weight	3.3–3.7kg (7¼–8⅛lb)	2.6–2.9kg (5¾–6⅓lb)	3.2–3.6kg (7–8lb)	1.2–2kg (2⅔–4⅓lb)
Body length	55cm (22in)	50cm (20in)	55cm (22in)	Up to 40cm (16in)
Ear length	10cm (4in)	7cm (2¾in)	7.5cm (3in)	6.5–7.5cm (2½–3in)
Hind foot length	14.5cm (5¾in)	14cm (5½in)	15.5cm (2⅛in)	7.5–10cm (3–4in)
Top of tail	Black	White	White	Black
Colour	Reddish brown, grizzled	Sandy or greyish brown, usually white in winter	Sandy or greyish brown	Greyish brown, variable
Gestation period	41–42 days	50 days	49 days	30 days
Young per female per year	Around 10	Around 6	Unknown	10 to >30
Young (at birth)	Furry and ready to run or hide, eyes open	Furry and ready to run or hide, eyes open	Furry and ready to run or hide, eyes open	Naked and not really mobile, eyes closed
Lifespan	2–3 years, occasionally up to 12 years	3–4 years, occasionally up to 18 years	3–4 years	Up to 2 years

Evolution and Adaptations

Evolution has taken place over such a long time-scale and under such changing circumstances that it can be difficult to comprehend. The hares that exist today, like other mammals, evolved as a result of constantly changing environmental pressures over millions of years. They were faced with melting ice sheets, and changing sea levels, vegetation, and predators, but they survived. Hares became fast, open habitat specialists with special adaptations that help them to avoid predation.

Depending on exactly how you define mammals, they evolved around 170–225 million years ago. Mammals produce milk to feed their young and have special distinguishing features in their brains and bones; most mammals also have body hair.

Fossils suggest that the rabbit-like mammals separated from the rodent-like ones more than 50 million years ago in Asia, though the rabbit-like mammals were classified as rodents until 1912. Within the mammalian order Lagomorpha (rabbits, hares and pikas – mammalian herbivores with four incisors in their upper jaws), the Leporidae family (rabbits and hares, defined by their upright head posture and strong hind limbs) evolved about 53 million years ago in Asia, and spread into Eurasia about eight million years ago. The genus *Lepus* (hares) is first recognised in the fossil record about 2.6 million years ago in the northern hemisphere (north of the Tropic of Cancer).

Opposite: Mountain Hares were common during the last glacial period and evolved in open tundra. Today they inhabit heather moorland.

Below: This fossilised skull of an extinct species of lagomorph (*Paleolagus haydeni*) from North America looks similar to skulls of modern lagomorphs (see page 18).

Hares during the Ice Age

Fossils that have been collected in caves, potholes and muddy lakes suggest that Mountain Hares were probably already in the British Isles during the Wolstonian glaciation, which started around 352,000 years ago and ended 130,000 years ago. Archaeological remains found in what is now Belgium suggest that Mountain Hares were eaten by humans 28,000 years ago. But to understand the mammals found in Europe today, including hares, we have to look to more recent evolutionary history, since the maximum extent of ice during the last glacial period (the glacial maximum) about 20,000 years ago.

At that time, Great Britain was not an island; southern England, southern Ireland and continental Europe were connected by dry land, as sea levels were lower than today. Northern Britain was connected to Denmark, but both were covered with ice sheets. Ice covered northern Britain as far south as present-day North Yorkshire, down the east coast towards Norfolk, and extended in the west to South Wales and southern Ireland. An ice cap covered the mountains of southern Ireland; only a small part of Ireland

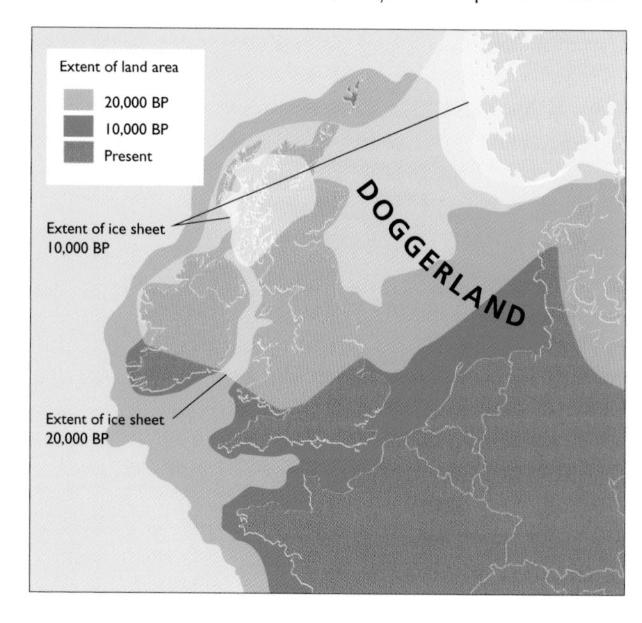

Right: Since the ice was at its maximum extent during the last glacial period 20,000 years ago (indicated by BP, before present), rising sea levels have resulted in changing coastlines in Europe. The retreating ice sheets were slowly replaced by forests, which were eventually cleared by humans.

remained ice-free. South of the ice sheet was bare tundra vegetation, where Mountain Hares probably felt quite at home – it was similar to the habitat they occupy today, and the habitat they evolved in.

Around 20,000 years ago, Mountain Hares were probably the most common and widespread hares in Europe, and their fossils have been found as far south as northern Spain. The Mountain Hare is one of a small handful of species still found in Britain today that could have survived the cold, harsh conditions here at the time. Other, more exotic species were also still present: Woolly Mammoths, Reindeer, Arctic Foxes, Woolly Rhinoceros, Lynx and Wild Horses. Fossilised bones from caves in England have cut marks showing that humans hunted Mountain Hares, for their fur or meat, approximately 12,000 years ago.

Melting ice sheets

After the retreat of the ice that began about 12,000 years ago, Europe slowly became covered in forest and so less suitable for hares. Mountain Hares gradually became restricted to areas with suitable habitats, and eventually only small, isolated relict populations remained in the Highlands of Scotland, parts of Ireland and the European Alps. The species' main population developed from Scandinavia to eastern Siberia in the wake of the retreating ice sheet.

The small populations of Mountain Hares that remained in Ireland and Scotland have been isolated from one another since long before the retreat of the ice and, as mentioned on page 7, may be in the process of becoming separate species. This happens by natural selection. Those individuals most suited to life in Ireland and best able to survive and breed there are different from those most suited to life in Scotland. Conditions for the two populations are not the same, so eventually, over many generations, the hares end up being different enough to be considered separate species. On the Scottish island of Mull, Irish Hares were introduced in 1863, and Scottish Mountain Hares were released in 1864. The two groups are believed to

Below: Mountain Hares in isolated relict populations may be becoming separate species.

Below: Two Irish Hares (foreground, and back right) with a possible cross between an Irish Hare and a Brown Hare (back left). The possible hybrid has a grizzled coat typical of a Brown Hare, but in size and shape it is more like an Irish Hare.

have remained separate for at least 30 years, suggesting that they were (and are) different species.

The Brown Hare evolved as an animal of open steppe habitats. It probably survived the harsh conditions of the last glacial maximum in the area around the Black Sea, and spread through Europe during the last 12,000 years as humans turned forest into more open agricultural land. Brown Hare may have been introduced to Britain before Roman times, but it can be difficult to distinguish its archaeological remains from those of the Mountain Hare as skeletons are often damaged or incomplete. Older remains may belong to the Mountain Hare, which was already present in Britain. Therefore, it is not entirely clear whether the Brown Hare was introduced to Britain or made its own way across Doggerland, the land bridge that existed between England and Denmark until about 8,000 years ago.

Interbreeding

Hybrids may be born as a result of breeding between two animals or plants of different species (or genera, subspecies, etc.). Hybridisation between species is confusing – by definition it cannot occur, because species are often defined as groups of animals that can only breed with other members of the same group. Most of the time, members of different species do not interbreed, because speciation (the evolution of new species) has led to mechanisms to put them off each other. Members of different species don't find each other attractive, their courtship behaviour is different, they smell 'wrong', the females' eggs cannot be fertilised by the sperm of the other species, or the males' penises are the wrong size or shape. Some taxonomists seem to be obsessed by measuring penises or penis bones (bacula), because differences in penis shape between species are often used to classify animals as belonging to different species, and taxonomists, like many other scientists, like working on things that can be measured.

Yet despite the preventative mechanisms, members of different species do sometimes mate. Often the offspring don't survive or are infertile, but occasionally they are healthy, active and even able to breed themselves. This is most likely when the hybridising parents are closely related, such as with species of the same genus. In some cases (including in *Lepus*), hybridisation occurs mainly in one direction – males of one species mate with females of another – and happens less often the other way around. This is often because only big, strong males get to mate, while all females are able to mate: females are selective and prefer big males as mates. Therefore, the bigger species has an advantage as a male, but not as a female.

Above: A cross between a Lion and a Tiger, sometimes called a liger or a tigon.

Above: A cross between a Peregrine Falcon and a Merlin, sometimes called a perlin.

Hybridising hares

Brown Hares and Mountain Hares now mostly live in different habitats but, where they are found together, they sometimes interbreed (or hybridise). Since the two species are closely related, their offspring may be fertile.

Hybridisation can be very harmful, and may even lead to the local extinction of the Mountain Hare. In southern Sweden, since the introduction of the Brown Hare in the 1800s, the native Mountain Hare has become extinct over a huge area. Changes in habitat and climate may have been detrimenwtal for the Mountain Hare. However, Brown Hares are also bigger than Mountain Hares and are able to keep Mountain Hares away from food. In addition, male Brown Hares, being bigger than their rivals, are able to mate with female Mountain Hares; any female hybrid offspring also tend to mate with big Brown Hares. After a few generations, the animals look and behave like Brown Hares but have traces of Mountain Hare DNA to demonstrate their history of interbreeding. Eventually, few 'pure' Mountain Hares remain.

In Ireland, around 15 separate introductions of the Brown Hare were recorded between 1848 and 1892. Though many populations died out, the non-native species became established at two locations in Northern Ireland. In one of these locations, about half the hares are now Brown Hares, and the area occupied by Brown Hares in Northern Ireland increased substantially between 2005 and 2013. As in Sweden, they compete and hybridise with the native hares. Brown Hares constitute a serious and increasing threat to Irish Hares, and there are even calls to eradicate them from Ireland.

Traces of Mountain Hare DNA, providing evidence of past hybridisation, have been found in Brown Hares from various parts of Europe – the Iberian Peninsula, Russia and Sweden so far. Thus a genetic legacy from an Arctic species reveals to us its wider distribution thousands of years ago, when the climate was colder than today.

What makes a hare?

Right: Blood vessels in the ears of a hare allow them to be used like car radiators, to cool the hare down.

Natural selection in countless generations of hares over the 6.2 million years of *Lepus* evolution has allowed them to become experts at surviving in open habitats, despite being attractive food for many predators. Some obvious and defining features of hares, essential to their success, are their long ears and back legs, and their excellent all-round eyesight.

Their long swivelling ears are multifunctional: they are used for acute directional hearing; as heat exchangers, allowing hot hares to cool down after running or in hot environments; as signals, to convey information to other hares, especially at dusk when their markings show up well; and as shock absorbers, to support the skull when hares are running at top speed. White markings on the tail also show up in low light and are used as social signals. Strong muscles in the large, extra-long back legs are

Above: Listening, looking and running: this hare is making good use of its senses.

Below: Running hares can lift their tails to signal with the white markings. Their ears work as shock absorbers.

adapted for powerful running, but also for fast and deft changes of direction: hares can jump, twist and turn to avoid capture. Brown Hares, when chased, are said to be able to run at 72km/h (45 miles/hour), while Mountain Hares can reach speeds of up to 64km/h (40 miles/hour).

Hares have excellent all-round eyesight, enabling them to spot predators. Using their ears as shock absorbers allows them to see well while moving at speed over the terrain: their eyes are stabilised by their ears streaming out behind them as they run. Like many prey animals, hares have eyes on the sides of their heads, so that they can see behind them as well as to the side and in front. Hares have a keen sense of smell and use odour in communication. They also have a good sense of touch via their long whiskers; they use them when moving around to gauge the size of gaps in vegetation.

Right: Hares have keen senses. Their eyes, like those of other prey animals, are positioned on the side of the head, allowing all-round vision. Their whiskers give them a good sense of touch.

Below: The contrasting markings on the backs of hares' ears show up well, even at dusk, and are used as signals.

Above: Even while they're grooming or scratching, hares are still checking for predators.

Below: During the moult, in spring and autumn, hares can look scruffy. Patches of fur differing in length are particularly obvious on the head, as the fur is relatively short there. This Brown Hare is starting to moult. From the head, the moult will progress down and back to the rest of the body.

Relatives Around the World

Hares, rabbits, and pikas all belong to the lagomorphs: four-legged, medium-sized mammals that biologists define by special features of their teeth. These close relatives are found on every continent except Antarctica, in many different habitats, including steppe, desert, forest, grassland, moor, scrub, coastal areas, mountains, agricultural land and parks. Lagomorphs are almost entirely herbivorous, unlike their closest relatives among the mammals, the rodents. In common with many rodents, lagomorphs are attractive prey items for lots of predators, and so they have mostly evolved to be experts at watching for danger, avoiding detection, running and hiding.

How we classify living things

Our ancestors would have been familiar with the animals and plants around them, and surely named them, but there were few attempts to standardise the names until Swedish botanist Carl Linnaeus started his life's work on taxonomy, the classification of living things. Along with thousands of other species of animal and plant, he described the Mountain Hare, and named it *Lepus timidus,* in 1758. Today, living things are classified into kingdoms, phyla, classes, orders, families, genera and species, though the classification is constantly being changed due to new discoveries. The names of genera and species are italicised and form the scientific name of each species as proposed by Linnaeus. Scientific names are useful to avoid any confusion about a species, particularly between scientists from different countries.

Above: Carl Linnaeus, the father of taxonomy.

Opposite: The Arctic Hare or Polar Rabbit (*Lepus arcticus*) is closely related to the Mountain Hare. This one, in Greenland, is standing on its hind legs to get a better view.

At international conferences, biologists usually use scientific names to refer to their study species.

The three closely related groups of lagomorphs differ greatly in size: with a few exceptions, pikas are tiny, weighing approximately 75–290g (2⅔–10oz), rabbits and cottontails are medium-sized (1–4kg; 2¼–8¾lb), and hares are bigger (2–5kg; 4⅓–11lb). Of the 87 lagomorph species that have been classified by the IUCN, 18 (three pikas, five hares and 10 rabbits) are considered to be Vulnerable, Endangered or Critically Endangered. Confusingly, some rabbits have common names including the word 'hare' (the Hispid Hare is really a rabbit), and some hares have common names including the word 'rabbit' (the Polar Rabbit and all jackrabbits are hares). The muddle exists only when common rather than scientific names are used – all members of the genus *Lepus*, and only those 32 species, are hares.

Scientific classification of hares

Kingdom	Animalia (animals)	
Phylum	Chordata (animals with a nerve cord)	
Class	Mammalia (mammals; animals that feed their young with milk)	
Order	Lagomorpha (approximately 91 species of rabbit, hare and pika)	
Family	Leporidae (rabbits, cottontails – around 29 species; and hares – 32 species; the other lagomorph family Ochotonidae contains the pikas, around 30 species)	
Genus	*Lepus* (hares and jackrabbits)	
Species	*timidus* (common name: Mountain Hare)	*europaeus* (common name: Brown or European Hare; above)
Subspecies	*scoticus* (Scotland) *hibernicus* (Ireland) plus around 13 others	*occidentalis* (England) plus around 15 others

A confusing local family

Six species of hare are found in Europe: the Corsican Hare and Broom Hare occupy small and separate ranges, whereas the Iberian Hare, the Brown Hare and the Mountain Hare have much wider ranges that overlap in some places. European Rabbits are also common and widespread in much of Europe. The hares found on Sardinia are believed to be introduced Cape Hares (*Lepus capensis*; also found in Africa, Arabia and northern India), though they may actually be Iberian Hares.

The Corsican Hare, Broom Hare, Cape Hare and Iberian Hare were all, at times, considered to be subspecies of the Brown Hare, and all look similar. The Corsican Hare and Broom Hare may be the same species, and some taxonomists believe that the Brown Hare is a subspecies of the Cape Hare. In fact, all the hare species in Europe can probably hybridise. It is confusing, and more research on the relationships between the hares in Europe is needed.

The Corsican Hare (*Lepus corsicanus*) is found south of the Abruzzo Mountains in Italy, on Sicily and in small numbers on Corsica, where it was introduced by humans, probably before 1400. It inhabits maquis (evergreen)

Below: Distributions of hares and rabbits in Europe overlap in most areas. The rabbits in the south-western Iberian Peninsula may belong to a different species, and it is unknown whether the Brown Hare is introduced or native in Britain.

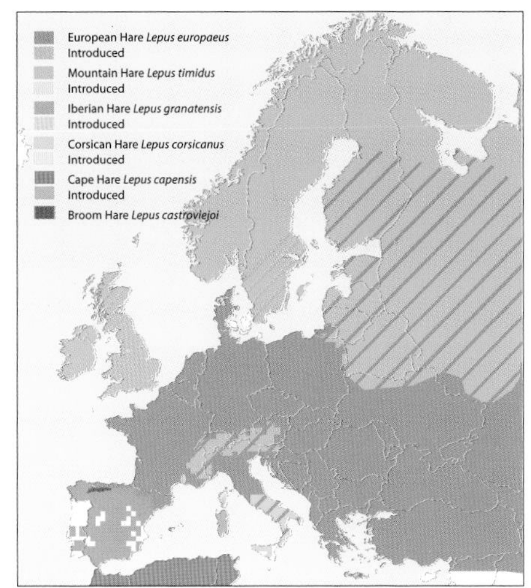

Hares

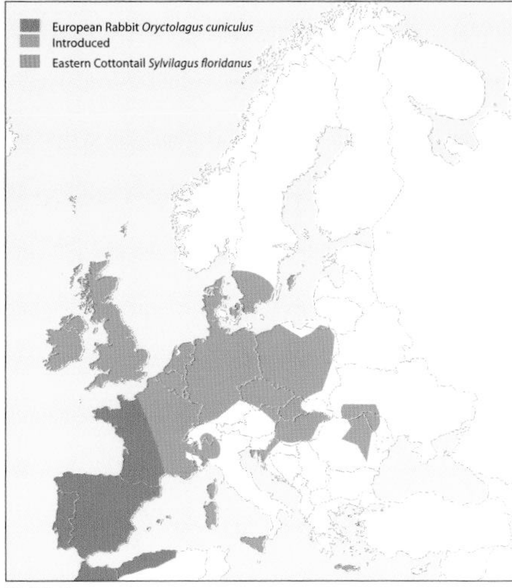

Rabbits

Above: The Cape Hare: introduced to Sardinia, naturally found in Africa, Arabia and northern India. In its native range, it is predated upon by the Cheetah, the only predator that can outrun the Cape Hare.

Right: The Corsican Hare was introduced to Corsica by humans, and a few are still there, but its stronghold is in southern Italy and on Sicily. It is considered Vulnerable by the IUCN.

shrubland, grassland, cultivated areas and dunes. The Brown Hare and the Iberian Hare were introduced to Corsica much more recently, in their thousands, and all three species hybridise there. The Corsican Hare's head and body length is 44–61cm (17–24in), body weight is 1.8–3.8kg (4–8⅓lb) and ear length is 9–13cm (3½–5⅛in) so it is smaller than the Brown Hare. The Corsican Hare is classed as Vulnerable by the IUCN; it is threatened by habitat loss, illegal hunting and competition with European Rabbits and introduced hares.

Without an expert guide, it would be almost impossible to see a Broom Hare (*Lepus castroviejoi*). This species is found only in the Cantabrian Mountains in northern Spain, in a region about 230km (143 miles) from east to west and 25–40km (16–25 miles) from north to south. It lives at elevations up to 2,000m (6,560ft),

Above: The Iberian Hare has a bright white underside that extends forward around the front legs.

Left: The Broom Hare, found only in the Cantabrian Mountains in northern Spain, is threatened by hunting.

though it moves down the mountains into warmer areas in winter. It is found in broom and heather heathland, and in clearings in oak and beech forests. The Broom Hare was not classified as separate from the Brown Hare until 1976, and little is known about its habits. It is classed as Vulnerable by the IUCN and the main threat to the species is unsustainable hunting of its small population.

The Iberian Hare or Granada Hare (*Lepus granatensis*) is found in the southern Iberian Peninsula, on Mallorca and in southern France, so it overlaps in places with the Brown Hare and with the Broom Hare. Its fur colour changes abruptly from grizzled brown to white on its sides, giving it a brighter appearance than the Brown Hare. It breeds all year round and occupies diverse habitats including arable land, mountainous forests and sand dunes; its numbers are believed to be stable.

Hares of snow, desert and water

Above: This Snowshoe Hare is moulting. Its large feet allow it to walk on snow without sinking into it.

Above right: Despite its winter coat camouflaging it in the snow, this Snowshoe Hare has been caught by a Canadian Lynx.

Lagomorphs are found in some surprising places. With special adaptations, they can thrive in cold, arid, and wet places. Large snow-shoe like feet have evolved in the Snowshoe Hare or Snowshoe Rabbit (*Lepus americanus*), the smallest species of hare (900g to 2kg/ 2–4⅓lb); it lives in the forests of northern North America. It is similar in size to a European Rabbit and has short ears to conserve energy by minimising heat loss. Its large, furry hind feet allow it to run on top of snow without sinking into it. It has extremely thick, insulating fur, which is white in winter (except for its black ear tips) and brown in summer. Climate change is likely to cause problems for this species: as snowfall decreases, the white winter coat will be less effective as camouflage. The Canadian Lynx, Coyote and Fisher are its main predators. Populations of Lynx and Snowshoe Hares follow predictable patterns – regular climatic changes cause rises and falls in the Snowshoe Hare population, which are followed two years later by rises and falls in the Lynx population. The population cycles last 8–11 years, and were first noticed over 200 years ago by trappers working in the fur trade.

The Tehuantepec Jackrabbit (*Lepus flavigularis*) is the rarest hare species. It is classed as Endangered by the IUCN and threatened by hunting and habitat loss. Fewer than 1,000 Tehuantepec Jackrabbits still survive today. It has the most southerly range of any hare species in North America, occurring in southern Mexico, around the Gulf of Tehuantepec. It inhabits grassland with scattered shrubs and trees, shrub forest and coastal grassy dunes within 4–5km (2½–3⅛ miles) of the shores of saltwater lagoons, and is not found at elevations over 500m (1,640ft).

Most hares need a regular source of water, but the Antelope Jackrabbit (*Lepus alleni*), found from the desert plains of south-central Arizona to the west coast of Mexico, prefers extremely dry desert scrub and does not need to

Above: An Endangered Tehuantepec Jackrabbit rests in its form during the day. It has large ears, and a characteristic black stripe (just visible) running from the base of its ear towards its back.

Left: The Antelope Jackrabbit is perfectly adapted to life in the desert. It uses its huge ears, which contain many blood vessels, to cool itself down.

drink water (it gets water mostly from the grasses, cacti and other plants it eats). It is large (3–4.5kg/6⅔–10lb), and has enormous ears (over 10cm/4in wide and 21cm/8¼in long) with very sparse hair. The ears are used to regulate body temperature – when it wants to cool down, the hare pumps blood into them. Females sometimes line their forms with fur and grass before they give birth. This species runs fast, and can leap several metres into the air. It sometimes stands on its hind feet; this allows it to feed on tall plants and to hop like a kangaroo for a few strides at a time. When running away from danger, the Antelope Jackrabbit displays a large white patch on its rump, which looks like the white rump of a Pronghorn Antelope, and gives the Antelope Jackrabbit its common name.

Most mammals can swim if they need to, but only one lagomorph makes a habit of swimming. The Swamp Rabbit (*Sylvilagus aquaticus*) is semi-aquatic. It is found in swamps and wetlands in south-central USA, and weighs 2–3kg (4⅓–6⅔lb). It swims to avoid predators, and may also lie still in the water, hidden by floating twigs and plants, with only its nose visible. It is hunted by humans, American Alligators and domestic dogs. It does not make burrows, but uses hare-like forms or nests.

Below: The Swamp Rabbit, the only lagomorph that regularly swims, has the American Alligator as one of its predators.

From superabundant to extremely rare

The European Rabbit (*Oryctolagus cuniculus*) probably has the widest range of the lagomorphs because it has been introduced in many places and is domesticated. It is a pest species outside its natural range (Spain, Portugal and north-western Africa), but within this range it is classed as Near Threatened by the IUCN. The rabbits in the south-western Iberian Peninsula may actually belong to a separate species, which has yet to be formally described. European Rabbits live in colonies and create large burrow systems (warrens) to avoid predation. Rabbits have aggressive encounters to establish a hierarchy – dominant males are able to mate with many females and so father more young. The European Rabbit is a key prey species that is critical for the conservation of species such as the Iberian Lynx and Imperial Eagle. Myxomatosis and Rabbit Haemorrhagic Disease have caused huge declines in Rabbit numbers, and Rabbits are also threatened by unsustainable hunting in their natural range.

The Riverine Rabbit (*Bunolagus monticularis*) is one of the rarest species of mammal: there are only about 250 adults, numbers are decreasing, and it is classed as Critically Endangered by the IUCN. It is found only in the central and southern parts of the Karoo Desert, in Cape Province, South Africa, in river basins and shrubland. It has a clear black stripe running from the corner of its mouth over its cheek, cream fur on its belly and throat, and a white ring around the eye; it weighs 1.4–2kg (3⅛–4⅓lb) and creates burrows. Female Riverine Rabbits produce only one young per year.

Below: A young European Rabbit. This species has become established in many places around the world.

Right: A Riverine Rabbit caught by an automated camera used as a survey tool by the Endangered Wildlife Trust in South Africa. The black stripe on the cheek and the cream fur on the throat are key identification features.

The striking Annamite Striped Rabbit (*Nesolagus timminsi*) inhabits evergreen rainforests in the Annamite Mountains on the border between Vietnam and Laos. It has seven clear black or dark brown stripes on its body and a reddish rump. It was first described in 1995 and first captured by biologists in 2015. Numbers are believed to be decreasing: the Annamite Striped Rabbit is hunted, and its habitat is threatened by logging and agricultural practices. It has a very restricted range and is almost certainly extremely rare and threatened.

Right: An Annamite Striped Rabbit caught on an automated camera used by researchers in the Quang Nam Saola Nature Reserve, Vietnam. The eyeshine that makes hares stand out in spotlight counts is visible here.

Tiny rabbits in special places

Hares and rabbits come in a range of sizes. At one end of the scale, we have the Brown Hare and the Arctic Hare, which weigh up to 5kg (11lb). At the other end, the smallest member of the Leporidae family is the Pygmy Rabbit (*Brachylagus idahoensis*), which is 25–29cm (9¾–11in) long and weighs just 300–400g (11–14oz). This tiny rabbit would easily fit in the palm of your hand. The Pygmy Rabbit is found in the parts of the western USA where its food plant, Big Sagebrush, occurs. It is the only native rabbit species in North America that digs. Most hares and rabbits are silent unless they are in extreme danger, but Pygmy Rabbits squeal, squeak and chuckle. There can't be much meat on a Pygmy Rabbit, but they are hunted by humans as well as by weasels, American Badgers, Coyotes, Bobcats, Red Foxes, and by birds of prey, both during the day (falcons) and at night (owls).

The Endangered Volcano Rabbit (*Romerolagus diazi*) is also very tiny (27–36cm/11–14in, 400–600g/14–21oz). This species is restricted to four volcanoes in the central Transverse Neovolcanic Belt in Mexico, and lives in a specific subalpine habitat containing bunchgrass (grass that grows in tussocks, rather than as a flat 'lawn'), and in some pine forests. It has very short ears and legs, and a minuscule tail. Only a few thousand individuals survive, their populations are fragmented and they are hunted for food by humans. It is also threatened by climate change, livestock grazing, agriculture, property development, logging, bunchgrass harvesting and forest fires. The future is very uncertain for this endangered species, especially because it can live only in a very specific subalpine habitat. Protection of the remaining individuals is urgently needed.

Above: This is the smallest member of the Leporidae family, the Pygmy Rabbit. Fourteen of these tiny rabbits weigh as much as one Brown Hare.

Below: The Volcano Rabbit is extremely tiny and also very rare. It is found only in a small part of Mexico.

The pikas

Pikas, also known as mouse-hares, rock-rabbits, hay-makers and whistling hares, are much smaller than the Leporidae (rabbits, cottontails and hares). The 30 or so species, all in the genus *Ochotona*, look like rodents because they lack the long ears of rabbits and hares. But they are better described as rabbits with egg-shaped bodies, short ears and apparently no tail (they actually have a relatively long tail, but it is never visible). They inhabit cold places, mostly mountains in central Asia and North America. Around 23 species are found in Asia.

Most pikas are more active during the day than at night. Pikas store food for the winter in 'hay piles', usually under rocks, which they defend aggressively against theft. The thieves are not only members of their own species; hay piles are also eaten by other hungry herbivores such as domestic cows and horses, Reindeer, goats, hares, marmots and voles. Mongolian herdsmen prefer to graze their animals where there are pikas because of the extra food provided by hay piles.

Pikas are much noisier than rabbits and hares: they whistle, squeak and 'sing' to communicate, and groups have distinct dialects. Some solitary, territorial species live in rocky screes; other more social species live in burrow systems in high-elevation steppe habitat. Pikas

Right: The American Pika or Little Chief Hare stores food for the winter in 'hay piles' under rocks.

Left: The cute Ili Pika is found only in northwest China. Its numbers are declining and it is classed as Endangered.

are important prey species, and in Russia they were killed in their thousands every year for their skins until the 1950s. The fur was used to make felt.

One of the largest but least-known species, the Endangered Ili Pika (*Ochotona iliensis*) was discovered in 1983 and has rarely been seen since. In the 1990s, there were only an estimated 2,000 individuals, and declines have occurred since then. The Ili Pika lives only in rocky screes on high cliff faces in the Tian Shan Mountains in Xinjiang, China. It weighs about 250g (8¾oz) and has brightly patterned, longish fur and rusty-red spots on its head; some people think it looks like a teddy bear. It is rarer, and arguably cuter, than the Giant Panda.

The American Pika (*Ochotona princeps*), also called the Little Chief Hare (the name used by the Chipewyan people), is found in high-elevation boulder fields in the mountains of western North America. It weighs 120–170g (4¼–6oz) and its body is 16–22cm (6⅓–8⅓in) long; it is greyish brown with white margins to its small, round ears, and long whiskers (4–8cm/1½–3⅛in). American Pikas often live under piles of broken rocks. They can only live in high, cool mountain regions and cannot tolerate high temperatures, so they may provide an early warning system to detect the effects of global warming – if numbers of American Pikas decline, temperatures may be increasing.

The Lives of Hares

You might think that hares spend their lives calmly munching grass at night and quietly resting, alone, during the day, and you would be right. However, underlying their timid, peaceful behaviour is a 'live fast, die young' strategy that has evolved under constant pressure from predators.

Speedy lifestyles

Many large mammal species have evolved to live for a relatively long time, breed slowly and invest precious time and energy in caring for their few young, while small mammals have evolved to breed quickly and many times during their short lives. In fact, there is a 'fast–slow continuum' along which each species has evolved to have its own place.

At the slow end of the scale, we have the whales and elephants. For example, the North Atlantic Right Whale may live for up to 70 years. Females give birth to one calf at a time, starting at about 10 years of age, after a gestation period of one year, and then breed only every three to four years. Over her long lifetime, a female Right Whale has on average only five or six young. You can see why the recovery of this species since whaling was stopped in the 1930s has been extremely slow.

Opposite: The Brown Hare has a speedy lifestyle. Females produce on average 10 leverets per year, but only live for a couple of years.

Below: A female Right Whale with a calf. She will give birth only five or six times during her long life.

At the fast end of the scale, we find the mice and rats. For example, a female House Mouse can breed at just 30 days old, and the gestation period is 20 days. She gives birth to a litter of 3–14 babies, and soon becomes pregnant again, though she lives for, on average, only one year. Populations of mice are able to increase very rapidly, and they can therefore make the most of any changes in their environment and recover quickly from declines.

The position of each species on the 'fast–slow continuum' can be predicted from its body size: small animals live fast; large animals live more slowly. Hares and rabbits, for their size, breed unexpectedly quickly and don't live as long as expected, so, like rodents, they are able to recover quickly from any declines and exploit changing environments. They are able to live 'in the fast lane' because their food is mostly plentiful and readily available. Female Brown Hares can breed at three to seven months old, give birth to 1–12 (typically four) leverets per litter, become pregnant again soon after giving birth, and have up to five litters per year. The average number of leverets produced by each female Mountain Hare in Scotland each year is estimated to be six; for Brown Hares in England, this number is 10. Rabbits, as you might expect, also breed 'like rabbits', but their naked young are born after a shorter gestation period.

Baby hares and rabbits grow exceptionally fast, both in the womb and after they are born. However, many young hares do not survive to adulthood. The population size of Brown Hares varies more due to changes in adult survival rates than it does due to the survival of young hares. Each year, about half of all adult Brown Hares die, while many more young die before they breed. Survival depends on many factors, including weather conditions and the time of year of birth – leverets born in early spring are much more likely to survive their first winter than those born in late summer, and can sometimes breed before their first winter. Wild Brown Hares are estimated to live for, on average, two to three years; the oldest marked individual was at least 12 and a half years old. Most deaths occur in the hard winter months, especially in Mountain Hares.

Daily routines

Most herbivores, including hares, spend a lot of time feeding. For mainly nocturnal or crepuscular animals like the hare, the short summer nights present a challenge: how can they fit in enough feeding time? The answer, for hares, is to start early and finish late. In summer, Brown Hares often leave their forms a few hours before sunset, while in winter they have a lie in until up to one hour after sunset. In summer, Brown Hares return to their forms up to five hours after sunrise, while in winter they return before the sun comes up. Following a similar pattern, Mountain Hares are most likely to be active in daylight at the start and at the end of the long summer days. They are active for about 13 hours each night in winter, but only for eight to nine hours in spring and summer. Mountain Hares in Sweden spend up to half of their active time in daylight in summer.

During the day, hares often rest in their forms with their ears back and their eyes partly closed. They also spend time grooming, usually at the beginning of their daytime rest period. From their forms, hares keep a good look out for approaching predators and listen for them too. If predators or humans approach too closely, hares leave their forms and bound away, usually travelling uphill (easy when you have short front legs and long hind

Above: This Brown Hare made its form in a cereal stubble field, but will need to find young green plants – weeds, crops or grass – to eat. It may need to spend time and energy traveling to a suitable feeding site.

Below: In summer, the short nights make it hard for hares to fit in enough feeding time.

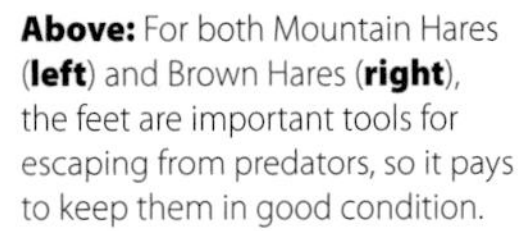

Above: For both Mountain Hares (**left**) and Brown Hares (**right**), the feet are important tools for escaping from predators, so it pays to keep them in good condition.

legs), and relying on their speed to avoid capture. The average distance at which an approaching human causes a Mountain Hare to leave its form has been measured in Scotland and in Finland; it is between 5m and 54m (16 and 177ft), depending on the air temperature. Strangely, hares seem to be more 'flighty' in colder weather. So, if you want to get close to a hare in its form, try to do so on a warm day – you should be able to get closer than on a cold day. It may be easiest to approach Brown Hares on wet days or when the ground is muddy – under these conditions they are not keen to leave their forms, because running in mud is hard work.

Right: Brown Hares tend to stay put in wet weather.

The main event – feeding

Left: Mountain Hares look around and swivel their ears while feeding, in case a predator approaches.

After they leave their forms in the evening, female hares may need to feed their young, but this only takes a few minutes (see page 66–68). In dry weather, hares sometimes roll in dust to clean themselves and remove parasites. Some time may be spent seeking mates, boxing potential mates or mating, but hares' main activity away from their forms is feeding. Hares are selective feeders and travel to the best areas where they can find the choicest morsels: the youngest, freshest plants and those with the highest energy, oil or sugar content. Scottish Mountain Hares usually travel downhill to feed, perhaps to make returning to their form easier (as long hind legs make for easier uphill travel). While feeding, hares are constantly looking and listening for predators and are ready to flee if necessary. The best fields may be covered with hares at night.

Groups of feeding Brown Hares of both sexes are sometimes structured so that the biggest and bossiest hares – those at the top of the hierarchy – get the best food. Once it has been established by boxing, maintaining the hierarchy requires almost no fighting, which suggests that, though hares are mostly solitary, they recognise each other and know who is boss. When they

Right: Brown Hares spend a lot of time eating grass.

form groups, Brown Hares are also able to spend more time feeding because they can share the responsibility of looking out for predators.

When returning to their forms at the end of the night, Brown Hares take an unpredictable route and sometimes make a couple of huge leaps, perhaps to baffle predators using scent to follow them. As additional protection, some hares have a few favourite forms that they use in rotation, moving every few days.

How many hares?

Counting hares to estimate the size of populations is difficult, but it has been attempted. Various methods have been used, including daytime walks to count hares in their forms, drives in which hares are flushed from their forms and counted, and counts of droppings or tracks, to derive what ecologists call an index of abundance. But counts made with different methods are difficult to compare, and there are added complications: were the counts made in spring, of the pre-breeding population, or in autumn, when numbers are swelled by the addition of youngsters, many of which may die during the winter?

Perhaps the best and most commonly used method is spotlight counting. Spotlight counts are conducted at night when hares are active. Researchers venture out armed with powerful torches and count the hares they see in the beam of light as they move it around in a circle from fixed points, or they point it into the darkness from a moving vehicle to create a strip of light. The eyes of hares, like those of many other mammals, have a reflective layer of tissue (the 'tapetum lucidum') behind the retina. This is an adaptation for life in low light – it allows the available light to be reflected back through the retina to maximise its use for vision. The tapetum lucidum also causes distinctive eyeshine, which helps hares to show up in spotlight counts.

From the numbers counted and an estimation of the area covered, or by using a mathematical estimation of visibility at different distances, the density of hares (numbers per sq km) can be calculated for different habitats or at different times. Researchers can also get an idea of where hares feed and whether they form groups or feed individually.

In most places where they have been counted, Mountain Hares reach densities of fewer than 10 individuals per sq km (26 per sq mile), but in some places, especially on islands, densities can reach a few hundred per sq km. In the optimum areas of north-east Scotland, densities may, exceptionally, reach 245 Mountain Hares per sq km (635 per sq mile). Densities of Brown Hares vary depending on the time of year and the habitat, but are on average 20–35 hares per sq km (around 52–91 per sq mile). One population on a small Danish island with few predators reached 200 hares per sq km (518 per sq mile), and populations in the best arable habitats may reach similar levels, especially in autumn. In some parts

Below: In most places it is unusual to see Brown Hares in groups. This photograph was taken on arable land in Germany, where Brown Hare density is high. It may show three males that are all interested in the same female as a potential mate, or a group of hares feeding.

Above: A researcher in Scotland using a radio antenna to locate a Mountain Hare that has been fitted with a radio transmitter.

Right: This Brown Hare, caught by Austrian researchers and fitted with a small ear tag for identification, is now ready for release. The blue bag was used to restrain the animal during handling.

of Europe, well over 20 Brown Hares per sq km (52 per sq mile) are shot each year.

Researchers tracking hares by radio telemetry found that their 'home ranges', the areas they occupy, are large and overlapping. Hares show no signs of being territorial, though as adults they do tend to stay in the same area. Mountain Hares in Scotland may travel a few kilometres to find suitable feeding grounds and, depending on the food available, may occupy an area of a few hectares (ha) to several square kilometres when they are feeding. Males usually travel slightly further than females and have bigger home ranges. For Brown Hares, home ranges are 20–190ha (49–470 acres), depending on the habitat.

Finding new places to live

If all the animals that were born into a population stayed where they were born, they may soon run out of food, and inbreeding would probably occur. To prevent this, animals disperse (permanently move away from their birthplace). Like young adult humans feeling the need to get away from their parents, other young animals, particularly males in mammals, instinctively move away and settle in a new area to breed. Female mammals also disperse, but are more likely than males to stay where they were born and eventually breed there.

Dispersal allows hares to colonise new areas and allows different populations to interact and interbreed; it has been described by ecologists as the glue that sticks local populations together. About half of all Brown Hares of four to six months old disperse, moving 1–17km (⅔–11 miles). There is much less evidence of dispersal in Mountain Hares; no dispersal is documented in Scotland. In Sweden, less than one third of young hares fitted with radio transmitters dispersed, and they did not go far.

Below: The snow is starting to melt around this Scottish Mountain Hare, and with perfect timing, its coat is moulting to match the landscape.

Breeding Machines

In winter in the northern hemisphere, while the days are short, changes start to take place inside the bodies of hares. Mating begins, and then for several months females are pregnant, suckling young, or both, while males are trying to find mates. The breeding season is a busy one for hares.

In winter, changes in day length act on the pineal gland in the brain and cause changes in the levels of a hormone, melatonin, that is only produced in darkness. Altered levels of melatonin in the blood act on male hares' genitals. The testicles enlarge, descend from the body cavity into the scrotum, and start to produce sperm, usually in December in male Brown Hares in England. In females, the uterus becomes ready for implantation of embryos around mid-January. Males start to sniff out females, and may follow them for hours, using scent to track them. Sometimes several males follow a single female.

If a male gets too close to a female before she is ready to mate, she strikes at him with her paws, boxing him away and testing his strength to evaluate him as a potential mate. In late December or January, females start to allow males to mate with them, and become pregnant with the first litter of the year.

Opposite: Brown Hares boxing. Females box males when they are not yet ready to mate, both to assess their strength and to chase them away.

Below: Hares have a keen sense of smell and use scent to identify other hares and to assess their interest in mating.

Busy breeding seasons

The mating season for Brown Hares in England is roughly January to September. Pregnant females are found in all months from January to August, and fertile males in all months except October and November. The breeding season for Mountain Hares in Scotland is shorter: males are fertile from January to June, and females are pregnant mainly from February to July or August. The gestation period in Mountain Hares is 50 days, as opposed to 41–42 days in Brown Hares. The longer gestation combined with a shorter breeding season and smaller litter size means that Mountain Hare mothers produce fewer leverets each year (circa six) than Brown Hares (10 in Europe). Mountain Hares have two or three litters each year (counted in Sweden), while Brown Hares have three to five.

In the northern hemisphere, a Brown Hare's first litter of the year is usually small – one, two or, occasionally three leverets – and is often born in February. Within a few hours of giving birth a female hare is ready to mate again, and if she finds a mate she may be pregnant and suckling her young at the same time for four weeks or so. In fact, she is likely to be pregnant, suckling, or both all summer until September. By March and April in England, female Brown Hares tend to be pregnant with bigger litters (three or more leverets). The biggest recorded litter of Brown Hares is 12 leverets. By the end of our summer (August), the last litter of the year is often small again.

Below: Male hares fight to establish a hierarchy, but boxing is most likely to happen when a female wants to show a male that she is not ready or willing to mate. She can test him at the same time. In this case the boxing is between two Scottish Mountain Hares.

Preparing to breed

Above: Against a snowy backdrop, well camouflaged Scottish Mountain Hares mate. It is early spring, the snow will soon start to melt, and this female is starting to get her summer brown coat.

In male Brown Hares, a mating hierarchy is established by biting and chasing. Once the hierarchy exists, there is little need for fighting between males. Like many mammals, hares use scent to identify themselves and others. Males rub their chins on vegetation in order to leave behind their own scent (from special glands). They also urinate on their feet and kick backwards to spread the smell, and lift their tails to release scent from glands. The bossiest males have access to the best food and guard females that are about to become ready to mate, keeping other males away. Male Scottish Mountain Hares are not known to guard potential mates, but dominant male Brown Hares stay within about 5m (16ft) of their target female, day and night. Dominant males are able to mate with more females and father more young than subordinate ones.

Above: Male Brown Hares guard a potential mate for several days before she is ready to mate, on the day that her eggs can be fertilised.

Female Brown Hares may be followed around by one or more males for up to five days before they are ready to mate. When they are nearly ready, they inform the males by shaking their tails and spreading a special scent. They allow males to mate with them only on the one day that their eggs can be fertilised. If they are not ready to mate, females turn away from males, drop their ears in a threatening posture, or box males that approach them closely.

The mad March hare - fact or myth?

Brown Hares have a long breeding season, so the popular belief that males in particular 'go mad' in March, fighting and chasing each other, seems strange. Why March, when male hares are fertile in other months? For many years, researchers have disagreed about this.

G. A. Lincoln, working in East Anglia in eastern England in the 1970s, believed that 'March madness' represented fights over females between testosterone-fuelled rutting male hares. He calculated that only 16 per cent of young hares were conceived during the period of 'March madness', and realised that Brown Hares have a long breeding season – most young are conceived in May, June and July. But he believed that males had high levels of testosterone in March, and that their fighting behaviour stimulated the females into breeding activity at the start of the season.

Tony Holley, watching Brown Hares in south-west England, saw that boxing actually occurs throughout the breeding season from February to September, but when the vegetation gets high and nights get longer this behaviour becomes harder to observe, hence boxing is most often seen in March. Tony also observed that almost all boxing takes place not between rival males, but between a female and a male. Female Brown Hares

Below: Hares don't just 'go mad' in March. Brown Hares can be seen boxing throughout their long breeding season.

Above: Boxing hares are quite aggressive – fur can fly.

are slightly bigger than males and, by boxing, are quite capable of showing they are not ready or willing to mate.

John Flux, researching hares in New Zealand used different terminology. He defined 'boxing' as fights between males to establish their hierarchy, and 'rebuffing' as females pushing males away when they are unwilling to mate. Like Lincoln, he believed that testosterone levels in males peak in March in the northern hemisphere (equivalent to September in New Zealand), and that males are aggressive at this time. He suggested that boxing (between males) is less common in populations in which the males all know each other and know their place in the hierarchy.

The jury is still out on the 'mad March hare', though it is clear that boxing has something to do with mating.

Boy or girl, old or young?

Male hares are called 'bucks' or 'jacks', females are called 'does' or 'jills'. There are many ways people claim to 'know' the sex of hares seen in the field. Some believe only females trail their ears behind them when running, others claim only males do that. Some people think only females run back through the line of beaters during hunting. Some think only males box, while others believe males and females 'flush' from their forms at different distances. The list goes on.

In fact, it is not easy to tell the sexes apart even when you have a dead hare in your hands, but it can be done by carefully examining the genitals. There is no truth in any of the claims above: male and female hares look very similar, and behave mostly in the same way. So, unless you are lucky enough to observe suckling, copulation or courtship, it is hard to tell the bucks from the does.

It is also far from easy to work out how old hares are. Pliny the Elder quotes the ancient Greek philosopher Archelaus as saying that a 'hare is as many years old as it has folds in the bowel', and also that hares are hermaphrodites. Both these claims are also, of course, untrue. Young hares are tricky to identify once they approach adult weight, but biologists use various methods.

The eye lenses of mammals continue to grow throughout their life, so the dry weight of the lenses can be used to work out the age of a hare. Extremely accurate scales are needed though – the eye lens of an adult Brown Hare, after drying, weighs a lot less than a small paperclip. The increase in eye lens weight slows as hares get older, so this method is most useful for assessing the age of hares that died very young.

Growth lines in the jaw bones can be used to determine the age in years of older animals; as with tree rings, a line on the bone forms each year. Complicated cleaning, drying, staining and sectioning methods are needed, and a microscope is used to see the lines.

As mammals reach adulthood, cartilage is replaced by bone in many parts of their bodies. A joint between bone and cartilage can be felt as a bump in one of the front leg bones of young Brown and Mountain Hares, up to about seven months old, but not in adults. This is the least labour-intensive method used to determine age, and the only one that can be carried out on a live hare, but it requires experience and is not always reliable.

Two litters at once

For centuries, scientists, philosophers and others have been debating the question of whether female Brown Hares can carry two litters of leverets at the same time, a phenomenon called superfoetation. Superfoetation in hares was described by the Greek philosopher Aristotle in the 4th century BC, and has been mentioned by many authors since. Pliny the Elder, at the time of the Roman Empire around AD 77, wrote: 'The hare which is born to be all creatures' prey is the only animal beside the shaggy-footed rabbit that practises superfoetation, rearing one leveret while at the same time carrying in the womb another clothed with hair and another bald and another still an embryo.' This was perhaps a slight exaggeration, as it suggests not just two, but three simultaneous litters at different stages of development, as well as one leveret still taking milk. There is, however, no doubt that hares can be simultaneously pregnant and feeding young.

Above: The Greek philosopher Aristotle wrote about hares 2,000 years ago.

Since the 1950s, there have been scientific studies of the phenomenon of superfoetation, and some evidence has been gathered. For female hares to be simultaneously pregnant with two litters at different stages of development, they would have to mate and become pregnant again when already pregnant. Normally, during pregnancy, eggs cannot be fertilised but superfoetation is believed to have occurred very occasionally in humans – women have given birth to 'twins' that seemed to have been conceived a month or so apart.

There is good evidence to suggest that superfoetation occurs in caged Brown Hares, where confinement of males and females together may result in more frequent mating. The second mating is likely to happen a few days before the female would normally give birth, and is evidenced by a gap between litters of less than the normal gestation period – around 36 days, instead of 42. Evidence also comes from ultrasound scans and dissection of pregnant females. The evidence from wild Brown Hares is less clear, but researchers have detected ovulation, suggesting the start of a second pregnancy, in three hares in late pregnancy. If fertile mating does commonly occur just before birth, it is unique among mammals, and it represents an adaptation for a 'fast' lifestyle, as it allows females to squeeze more pregnancies into each breeding season.

How many leverets?

To find out how many leverets each wild female hare has given birth to, biologists examine the inside lining of the womb in hares shot by hunters in late autumn. Each leveret leaves a placental scar when it is born. Conveniently, the scars fade over time, so each female hare examined in late autumn carries a record of her offspring for only the breeding season that has just finished. Placental scars occur in most mammals, including humans. For hares, the scar-counting method was developed in France. The scars can be grouped into litters: the most recently born leverets leave the clearest scars.

Knowing how many young are born to each female each year is very useful for researchers interested in population dynamics (how numbers of hares change over time), conservation and habitat management, and reproductive biology. With this knowledge, they can answer questions about how breeding is affected by the diet, body condition, health, age and size of the mother.

Right: This Brown Hare leveret could be a singleton or one of a large litter. In order to understand how numbers of hares change over time, ecologists need to know how many leverets are born to each female each year.

Above: The mother of these young Brown Hare leverets selected a sparse arable crop for their birthplace, and they are well camouflaged here. Even in grass, they are surprisingly hard to spot.

Below: Leverets grow very fast. These ones, all from the same litter, are probably waiting for their mother to feed them at dusk.

Growing Up Fast

From birth as tiny leverets to adulthood, hares are at particular risk from predators and infectious diseases, and their main task is to survive. They spend a lot of time hidden away in individual forms, though in the first few weeks of life they are cared for by their mothers. They have evolved to keep warm and safe, and to grow fast: within a few months, they reach adult size.

Adult hares are tempting prey for many predators and leverets are even better. They are perfect bite-sized morsels for animals such as Foxes, and the probability of a leveret ending up as Fox food (killed, or scavenged once dead) is high. They may also be attacked by Stoats, cats, Buzzards, Magpies and other predators, die of disease, or be killed by farm machines, cars or other vehicles.

A newborn Brown Hare leveret has about a 50 per cent chance of surviving to the end of the suckling period, and about a 30 per cent chance of surviving for the first year of its life; thereafter, if it is lucky, it has a 50 per cent chance of survival each year. It's not a great prognosis, and life for a leveret is mainly about avoiding predators. Without the protection of burrows, hares are particularly vulnerable but they have evolved ways to maximise their chances of survival to adulthood.

Opposite: The fluffy fur of a leveret is more useful for camouflage than for insulation.

Below: Brown Hare littermates separate soon after birth. Unless it is dusk, when they meet to feed, leverets found together like this are likely to be very young. These three all have white spots on their heads.

First few days

Above: Brown Hare leverets weigh about the same as an apple when they are born, but grow quickly.

At birth, Brown Hare leverets weigh only 100g (3½oz), – about the same as an apple. Mountain Hares in Scotland weigh about 90g (3⅛oz), and Irish Hares are believed to weigh 170–180g (6–6⅓oz), though few have been weighed. The weight depends on the number of leverets in the litter – bigger litters consist of smaller leverets, just as human twins typically each weigh less than a single baby would. Close observation of the behaviour of mother hares and leverets has only been carried out for the Brown Hare, however a similar pattern of vital but apparently minimal maternal care also occurs in Mountain Hares. Hares do not receive any care from their fathers.

Leverets are born in what seems a randomly chosen place, though no doubt the mother has her reasons for its selection; she spends a lot of time there in the few weeks before giving birth. The leverets are fully furred at birth and their eyes are open. They quickly move away from each other and find shelter. Some leverets are born with white spots on their heads, which may help to camouflage the leverets. It is unknown why only some leverets have them, and why they are not seen in adults. Leverets with white spots are – incorrectly – said to be males, or leverets from litters of a certain size.

Leverets have little odour and keep very still, each remaining separate in its own tiny form during the day. About 45 minutes after sunset each evening leveret littermates come together at their place of birth, where they play and run around for a few minutes. Eventually they sit still, perhaps grooming themselves and each other, and await their mother. She turns up about 15 minutes after they do, having spent the day a few hundred metres away. The leverets approach their mother; they will sometimes approach other passing adult hares, so the mother sniffs her leverets to check their identity. Then she may lick them to remove any urine or dirt on their fur that might make their scent stronger and give them away to predators.

In the first week after birth, the mother allows the leverets to drink her milk for up to five minutes each

Above: Once they have grown a little, leverets come together only to await their mother for feeding each evening.

Below: Older leverets run around to get to know their environment.

evening. Mother hares have six nipples, so with litters of more than six, leverets would have to take turns to feed. Each leveret feeds for only a few minutes and that is it for the whole day – their one quick feed has to give them enough energy to last 24 hours. After providing milk, the mother jumps away, leaving her leverets to their own devices until they meet again at the same time the next evening. If you ever come across a leveret in its own form, don't assume it has been abandoned and is in need of help from humans. It is probably experiencing the normal solitude of a leveret, and will meet its mother at dusk.

It seems a lonely existence, but this is how leverets avoid being eaten by Foxes and other predators. If the littermates spent all their time together and the group was discovered by a predator, they would all be eaten and the entire litter would perish. By spreading themselves out, the littermates make it more likely that at least one of them will survive to adulthood.

Suckling period

When they are a few days old, the leverets move further away from each other, and each end up about 100m (328ft) from their birthplace during the day. The time the mother allows them to feed for decreases, from five minutes in the first week to only one to two minutes as they get older, and suckling still takes place only once every 24 hours.

If the birthplace changes in character (perhaps through flooding or farming operations) and is no longer deemed suitable, the mother and her leverets search for each other and meet up somewhere else. If she is disturbed during suckling, the female leads the leverets away with her tail held up, the white tail-patch acting as a signal for her leverets to follow. Researchers noticed that when a Barn Owl came to a litter of leverets' birthplace at around the time they were due to suckle, the mother and leverets took cover until it had left, before suckling as usual later on. Mountain Hare mothers are very loyal to their chosen birthplace and they will return to it even after being disturbed.

Above: Brown Hare mothers take only a few minutes each day to feed their young.

Brown Hare mothers give their leverets milk for at least 17–23 days. They carry on for longer – often for 30 days, or occasionally for over 67 days – with big litters and their last litter of the season. Scottish Mountain Hare mothers provide milk for around 28 days, and will also continue for a little longer, up to 42 days or even more, for their last litter of the year. At the end of the breeding season, when she is no longer pregnant, the mother may feed her final litter of leverets for longer because it will help them to make it through the harsh winter, when food is less readily available.

Supermilk

How can the tiny leverets grow so fast when they are fed so infrequently? At 30 days of age, Brown Hare leverets weigh on average 8½ times their birth weight. This growth rate is phenomenal. Imagine the same growth rate for a human baby. The average birth weight for humans is 3.5kg (7¾lb); 8½ times that is 30kg (66lb) – the weight of a nine-year-old child! Scottish Mountain Hares also grow fast – 14g (½oz) per day in the wild, or 30g (1oz) per day in captivity.

Leverets do start to feed on grass and other vegetation from about day 15 of their life, so they are not totally dependent on milk. Leverets from large litters eat more solid food than those from smaller litters. But there must be something very special about hares' milk to allow the leverets to grow so fast while spending so little time drinking it. Scientists at the Research Institute of Wildlife Ecology in Vienna devised a hare-milking machine in order to collect milk from captive hares for analysis. They found that the milk consisted of 20–26 per cent fat. How does this compare to the fat content in other mammals' milk? Well, fat content is quite variable, from as low as 0.2 per cent in Black Rhinoceroses' milk up to 60 per cent in the milk of some seals that have to suckle their young for only a few days on constantly changing ice

Right: A mother Mountain Hare provides milk for her leveret.

Below: To analyse captive hares' milk, scientists devised a hare-milking machine.

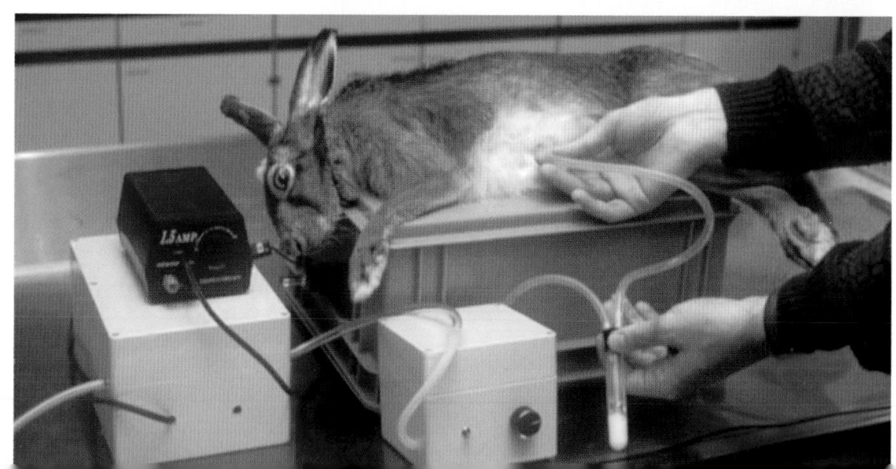

floes. Animals that suckle their young little and often, such as cows and primates, produce milk that is lower in fat than those that suckle rarely, such as hares. Cows' milk is 3–5 per cent fat and human milk is about 4.5 per cent.

The fat in hares' milk comes mainly from the mothers' food, and captive hares fed a high-calorie diet eat less but produce fattier milk than those fed on standard hare food. This explains why hares are such selective feeders: they need a high-energy diet to aid their escape from predators, which would be more difficult if they had to run with a bulky stomach full of low-energy food. But they also need to provide high-quality milk for their young, so they can feed them only briefly, once every 24 hours. Like so many of the hare's features, its parental behaviour has evolved to help it and its young avoid predation.

Tony Holley – hare-watcher

Making close observations of the private lives of secretive, well-hidden hares without disturbing them is a huge challenge, and few people have ever attempted it systematically. On the Somerset Levels in south-west England, Tony Holley became fascinated by the Brown Hares he could see from his house. He bought a small piece of grassland and managed it specifically for hares, allowing some cattle grazing to create the perfect mixture of grasses, reeds and clumpy thistles. Hares liked the conditions there and used it for feeding, breeding and resting.

Between 1977 and 1987, Tony spent thousands of hours observing hares with high-powered binoculars and telescopes. He was able to identify individual hares from natural markings on their bodies, such as small scars, ear nicks and fur patches. By occasionally glimpsing male genitals and female teats, he was also able to sex the hares. He was the first person to realise that male hares have a hierarchy relating not only to food but also to mates, and that dominant males are able to mate with more females than subordinate ones. He named the hares he observed and watched their most intimate moments, recording and photographing boxing, mating, feeding, mothers with leverets, and many other aspects of hares' lives. He studied interactions between hares and predators, including Foxes and crows, and he wrote about the timing of their activity, their use of forms, and about aspects of grooming, stretching and play in leverets.

Tony wrote his PhD thesis on hares and published papers that are still used by biologists. His observations were second to none, and much of what we know about Brown Hare behaviour comes from Tony.

Playing and exploring

Leverets spend time exploring their surroundings in their first few days of life. They need to know their way around in case of attack by a predator, or in case they are flushed from their form and need to get back to it. They also play. Brown Hare leverets often run conspicuously up and down a familiar path (called 'streaking' on the 'race track' by researcher Tony Holley). They sometimes hold their tails up while doing this, and occasionally jump around. Each leveret in a litter has its own separate 'race track', and they converge at the site where the littermates meet to feed each evening (their birthplace).

Surviving harsh weather

As well as avoiding detection and surviving attacks by Foxes and other predators, leverets have to keep themselves warm and dry. Imagine how cold you could get as a tiny leveret born in the open in February. Researchers in Austria found that, during the long breeding season there, Brown Hare leverets may be exposed to extremes of temperature (–26°C to +38°C/–15°F to +100°F), and monthly rainfall can be up to 244mm (9⅔in). They tested the ability of leverets in their first week of life to cope with low temperatures and found that, at least down to –8°C (18°F), leverets are able to cope with the cold by increasing the heat they produce via energy from food. Though leverets are furred at birth, their fur is mainly for camouflage and does not provide much insulation. It is likely that, when it's cold, leverets restrict the blood flow to their skin to minimise heat loss; they feel cold to the touch, but their internal body temperature remains stable.

Despite their adaptations to harsh weather conditions, leverets do need more energy when it is cold, and there is a price to pay for spending energy on keeping warm: slower body growth and higher likelihood of death from illness. Leverets exposed to lots of rain are less likely to survive than those experiencing drier conditions. Captive

Above: Young hares like this one need to eat more when the weather is wet or cold – their survival may depend on it..

young Brown Hares in dry, sheltered cages grow faster than wild ones do, which shows that wild hares need to spend lots of energy just on keeping warm. For wild leverets, shelter and food quality are important, which is perhaps why habitat management can significantly benefit populations of hares.

In sickness and in health

Wild hares can become infected by a range of viruses, bacteria, yeasts, internal parasites such as worms, and external parasites such as fleas and ticks. Some of these cause sickness and death, others probably make infected hares more likely to be shot, killed by a predator or run over by a car, while some seem to have no adverse effect on the hares.

Most illnesses affecting hares, as far as we know, are shared by Brown Hares and Mountain Hares. However, European Brown Hare Syndrome, caused by a type of virus, expressing itself as a form of viral hepatitis, affects mainly adult Brown Hares in autumn. It can cause death 5–24 hours after the first symptoms appear. The disease was identified in 14 per cent of a large sample of Brown

Right: Hares that are not in tip-top condition are the most likely to fall prey to Foxes or other predators.

Hares from Sweden and is also common in other European countries. In Croatia in 1986, 30,000 hares were estimated to have died of European Brown Hare Syndrome. The virus is transmitted via direct contact between hares, or when hares come into contact with infected droppings.

Pseudotuberculosis (caused by the bacterium *Yersinia pseudotuberculosis*) is found in up to 20 per cent of hares found dead. Mountain Hares can also be infected with the bacteria *Pasteurella, Leptospira interrogans* and *Francisella tularensis,* and 60 per cent of Brown Hares sampled in the Netherlands were infected with the bacterium that causes syphilis.

Coccidiosis (caused by *Eimeria,* a protozoan or single-celled animal-like organism) affects mainly young hares in autumn, and is perhaps the most common natural cause of death in hares in the British Isles. It is an infectious disease, probably transmitted when hares come into contact with other infected hares or with their droppings.

Helminth worm parasites are found in the livers, stomachs, lungs and intestines of the majority of wild hares. The stomach worm *Graphidium strigosum* is mainly found in Rabbits, but infects hares where Rabbits and hares live close together. External parasites found on hares include many species of fleas, lice, mites and ticks.

Hares occasionally suffer from grass sickness, a disease of the central nervous system that is associated with lush grazing and also occurs in horses. The cause of grass sickness is unknown.

Becoming an adult

Young hares can be born at any time during the long breeding season but, regardless of their age and size, in order to have any offspring they eventually have to be ready to mate in the mating season. The age at which hares reach puberty depends on when they were born, and their growth depends on the habitat and weather conditions.

In England, male Brown Hares born before May reach puberty at three months of age, so, if they get access to any females, they can breed in their year of birth. They are still quite small when they become fertile (2–2.5kg/4⅓–5½lb) and whether they actually father any young is unknown, but they are physically able to. Males born from May onwards don't become fertile until the beginning of the next breeding season, in December, by which time they may be seven months old. In warmer southern Europe, hares are more likely to be able to breed in the year of their birth. In Australia, female Brown Hares can breed at three months of age; in England, they can breed from around five months old, but this is rare and most first-time mothers are older.

As well as being able to breed, adults are defined as being of adult size and weight. Brown Hares reach their adult weight at about six months of age. Mountain Hare leverets are smaller than adults until they are about four months old, and, as far as we know, they don't breed in their year of birth.

Below: Leverets grow quickly and reach adult size after only a few months.

Eat or Be Eaten

Like most herbivores, hares appear to be surrounded by food that is much easier to obtain than the food of carnivores is. For herbivores, there is no need for tiring, dangerous and potentially unsuccessful hunting. But the apparent easy availability of food plants is deceptive. Hares' survival depends on them eating high-quality, high-energy food, and they must avoid eating large amounts of low-quality food that would fill them up but slow them down. Many predators are looking out for hares to catch and eat, and hares rely mainly on their speed to avoid being caught.

Food for hares

Brown Hares prefer to eat young, fresh shoots, and diversity is important in their diet, but their main food consists of grasses. They graze selectively on grasses, other herbaceous plants and arable crops (young cereals, in particular wheat, but also maize, peas, beans, sugar beet, and ears of cereals). In pastoral land, hares eat a greater diversity of plants than they do in mixed farmland; their diet is even less diverse in agricultural monocultures. In one study in Sweden, at least 37 different plant species

Opposite: This moulting Brown Hare is eating clover while keeping watch for predators.

Below: In cereal stubble, hares are likely be be feeding on weeds.

Right: A mixed agricultural landscape offers diverse foods and cover for hares at different times of year.

Right: Monocultures, where only one or very few crops are grown, are not beneficial to hares.

were eaten by Brown Hares in pastoral areas, 30 were eaten by hares in areas of mixed farmland, whereas only 14 were eaten by hares in monocultures. In an arable area of Austria, of the 164 plant species available to them, hares ate 49 and actively selected 11, including crops (soya beans and winter wheat), sugar beet and carrots provided by hunters, and wild plants (e.g. clover and poppy). However, hares in mixed farmland in north Yorkshire relied heavily on three grass species and wheat.

Wild grasses and herbs may be preferred to cereals and other crops, and in France, Field Horsetail is particularly popular in summer. Flowers are also eaten. Grasses form up to 90 per cent of Brown Hares' diet in

Left: This Mountain Hare is nibbling at the fresh shoots of heather, its main winter food. It is watching out for danger while it eats.

winter and are very important to them all year round. In summer, other herbaceous plants form 40–60 per cent of their diet. In winter in some places, cultivated plants are key to survival. Brown Hares living in conifer plantations eat mainly wild grasses, but they may also feed on shrubs and trees, and eat pine needles, particularly if snow makes grazing difficult.

Mountain Hares in Scotland have a slightly more 'woody' diet: they eat mostly heather in winter and grasses (their preferred food) in summer. They also eat clover, willow, rushes, sedges, Crowberry, birch, poplar, pine, Juniper, Dog Rose and lichens. They sometimes use their forepaws to dig through a layer of snow to reach

Right: This Mountain Hare's food, heather, is covered with a layer of snow, making feeding difficult and energetically costly.

heather, and often feed with their backs to the wind. Irish Hares eat mainly grasses; their diet is similar to that of the Brown Hare.

In terms of energy from their diet, wild Brown Hares in East Anglia, eastern England, are estimated to need 2,310kJ (552kcal) per day in summer, and 2,835kJ (678kcal) per day in winter. For comparison, men need to consume around 10,500 kJ (2,510kcal) and women 8,400 kJ (2,010kcal) daily to maintain a stable body weight. So a human, weighing on average about 20 times as much as a hare, needs about three or four times as much fuel as a hare. Hares are very active, and partly due to their small size, need a lot of energy to keep warm.

Below: Brown Hares are selective feeders, searching out the best quality plants.

How do we know what hares eat?

Hares are secretive animals that forage mostly at night, making close observation of their feeding habits difficult. Some researchers observe feeding hares or follow tracks made by hares in snow in order to identify what they have eaten, but most can only analyse stomach contents or droppings. The epidermis, or outer layer, of plants has distinctive microscopic structures which mean particular species or genera can be identified from fragments.

A sample of mushy, partially digested plant matter from a hare's stomach is dried and milled to produce plant fragments of an even size, or the sample is washed and boiled to produce transparent fragments. These fragments are then examined under a microscope and compared to a reference collection. To create a reference collection, plants are collected in a hare's foraging area. These plants are identified before being chopped into tiny fragments by researchers who are trying to mimic what happens when the plants are bitten off and chewed up by the hare. Reference microscope slides are made, and the features of these fragments are compared to those seen in the samples from the hare's stomach.

After many painstaking hours of study, plants in a hare's diet can be identified and the plant's abundance in its diet can be quantified. The diet can then be compared to the abundance of plants in the habitats used by the hare, to discover whether it is selecting or avoiding particular plants, or just eating them randomly in accordance with their abundance. Researchers have used this method to compare the diet of male and female hares, of hares at different times of year, and of hares in different habitats.

Eating everything twice

Herbivorous mammals have a problem – they cannot digest cellulose, the type of sugar that makes up plant cell walls. To solve the problem, herbivores have bacteria in their gut that digest the cellulose for them by a process known as fermentation. The bacteria convert the cellulose into fatty acids that can be absorbed by the herbivore.

In the 'foregut fermenters' (herbivores such as cows, sheep, deer and kangaroos), this fermentation happens before the food reaches the main stomach, either in separate chambers (extra stomachs) or in a modified gut. Some chew the cud – regurgitate partially fermented food and chew it again to improve digestion. In the 'hindgut fermenters' (such as elephants, horses, guinea pigs, wombats, rabbits and hares), fermentation takes place in the caecum, a chamber that comes after the stomach. Some, including the lagomorphs and a handful of rodent, marsupial and primate species, maximise their ability to use the fatty acids produced by bacteria during fermentation by eating their own faeces.

During the day, hares produce soft pellets called caecotrophs, made from food on its first round through the digestive tract; these are eaten – a process known as caecotrophy. Hard, round droppings, made from food on its second round, are produced at night. These are only eaten in times of food shortage. The soft caecotroph pellets are rich in protein and vitamins and by eating them, lagomorphs make use of nutrients missed during the food's first passage through the digestive tract. They are an essential part of the diet: domestic rabbits fed a normal diet but prevented from eating caecotrophs develop malnutrition. When feeding by caecotrophy, Brown Hares sit on their haunches and take pellets directly into their mouths from their anus.

Hares as food

It has been estimated that over five million Brown Hares are killed by humans in Europe each year. How this compares to the numbers taken by wild predators is unknown. Animals that eat hares are many and varied, and include mammals and birds of prey. Some are probably only able to take young hares, and some can only scavenge dead ones. For the Brown Hare in Europe and Australia, the Mountain Hare in Scotland and the Irish Hare, the Red Fox is the main predator.

Above: Brown Hares are game animals in most of their geographical range.

Below: The Red Fox is the principal predator of the Brown Hare.

It may be possible to tell if a dead hare has been eaten by a Fox, but it is impossible to tell whether that hare was actually killed by the Fox, by a car, by disease or by starvation. Thanks to some dedicated researchers who followed almost 800km (497 miles) of Fox tracks in the snow, we know that Foxes do feed on hares that they find dead. Foxes also hunt hares, though of 45 observed attempts at hunting, only three were successful. It must be hard being a Fox: the researchers discovered that to catch one hare, a Fox had to travel on average 263km (163 miles). Maybe only ill, starving or injured hares, or those struggling to run in deep snow, are slow enough to be caught. Leverets are perhaps easier to catch than adult hares.

Animals known to eat Brown, Mountain or Irish Hares throughout their ranges

Mammals	Birds
Scottish Wildcat	White-tailed Eagle
Domestic/feral cat	Eastern Imperial Eagle
Eurasian Lynx	Golden Eagle
Red Fox	Hen Harrier
Domestic/feral dog	Western Marsh Harrier
Wolf	Northern Goshawk
Arctic Fox	Rough-legged Buzzard
Badger	Common Buzzard
Wolverine	Ural Owl
Pine Marten	Snowy Owl
Beech Marten	Tawny Owl
Mink	Long-eared Owl
Stoat	Eagle Owl
Polecat	Black Kite
Brown Bear	Red Kite
	Peregrine Falcon
	Saker Falcon
	Raven
	Carrion Crow

Above: A Fox carrying a young Mountain Hare.

Above: Some important predators of hares, (clockwise from top left): Buzzard, Scottish Wildcat, Golden Eagle and Wolf.

Foxes eat hares all year round, but eat more in winter and spring than in summer and autumn, and more when other food (mainly voles and other rodents) is less common. The Brown Hare can constitute up to 50 per cent of the diet of Foxes. Other predators rely on Brown Hares less. The Buzzard is perhaps the main predator among the birds; though the Golden Eagle also relies on hares, it is more likely to have access to Mountain Hares (which may make up to 50 per cent of its diet) than to Brown Hares.

Foxes are opportunist feeders, so the extent to which they feed on hares is hugely variable. In Sweden, Mountain Hares make up 4–10 per cent of the winter diet of Red Foxes, and 1–14 per cent of the summer diet, depending on vole abundance. Some Foxes rely on Mountain Hares a lot more heavily; they can form up to 94 per cent of their diet. Wolves, Eurasian Lynx, Pine Martens and Arctic Foxes also rely on Mountain Hares at times. Rabbits and hares are consistently identified as the most significant element of the diet of Golden Eagles throughout their geographical range, and are therefore key to the survival of Golden Eagles.

Do predators make a difference?

The biology of hares has been shaped during evolution by constant pressure from predators, so populations of hares are likely to be able to cope with predation. However, numbers of hares may change with numbers of predators, and hares may become more or less susceptible to predation if their habitat changes. Understanding interactions between predators and prey is important for conservation and management.

It is difficult to know how predation affects hare populations, partly because many predators not only catch and kill hares, but also feed on carcasses, so they may be feeding on animals that died by other causes. However, in Scandinavia, when Fox numbers crashed due to disease, numbers of both Brown Hares and Mountain Hares increased. Predation by Foxes can slow down the growth of hare populations and reduce the density of hares; in England, it has been estimated that 80–100 per cent of leverets are taken by Foxes each year. In Sweden, 40 per cent of the leverets born each year are estimated to be taken by predators (mainly Foxes and cats). Mountain Hare numbers can decline when Foxes are around. Even in a part of Poland where Foxes were kept at low levels by gamekeepers, about 10 per cent of leverets were taken, and 18 per cent were taken in an area of higher Fox density.

Below: This Brown Hare is standing on its hind legs, perhaps to signal to a predator that it has been spotted.

It is believed that up to 50 per cent of adult hares are killed or eaten by Foxes. Researchers worked out that each Fox eats 30–45kg (66–99lb) of hare each year, suggesting that 10–15 per cent of all hares that die each year are eaten by Foxes. However, game bag data from Germany suggest that habitat improvement has more potential to benefit hares than does the removal of Foxes.

Avoiding being eaten

Above: Brown Hares can run faster in short vegetation, so they avoid tall plants when they feel threatened.

Hares sometimes swim to escape predators, and Mountain Hares hide in shrubby places when predators are near. Foxes trying to catch adult hares usually either approach them stealthily while they are resting, or attempt to outsprint them over short distances. Research in the Netherlands shows that when hares are approached by potential predators (in the study, a dog on a lead with two humans), they move faster and further when they are in areas of short vegetation than when they are in taller shrubby vegetation. Hares spent more time in taller vegetation for 24 hours after they saw the 'predators' – presumably hiding – so the effect of the disturbance is not short-lived.

Hares are clearly cautious animals. If you are a hare, the best way to avoid being eaten is probably to see the predator before it sees you, and be ready to run. When approached by a Fox or a cat in the open, hares sometimes stand on their hind legs facing the potential predator, to signal to the predator that it has been seen and therefore has little chance of catching them.

Conservation of Hares

There are many reasons for conserving hares, including: to ensure that they are not lost as a species; to help efforts to protect their natural predators, some of which are endangered species; to preserve biodiversity; and to ensure the future availability of hares as game animals. With their speedy lifestyles, fast breeding and abundant food, hares ought to be easy to conserve, yet many populations are declining. Brown Hares are also considered by some to be agricultural pests, especially in areas where they feed on soya beans, sugar beet and other crops, and young trees or shrubs, so not everyone is on their side.

Threats and declines

Native Irish Hares, like Mountain Hares in Sweden and the Alps, are threatened by expanding populations of invasive Brown Hares that compete and hybridise with them. Irish Hares are also believed to be declining in number due to habitat loss and other changes in the agricultural landscape that affect Brown Hares too. In western Scotland, Mountain Hares have already disappeared from some areas, probably due to the replacement of heather moorland, their favoured habitat, with forestry. Overall though, looking at the bigger picture, populations of Mountain Hares are stable.

Populations of Brown Hares in many parts of Europe have decreased since the 1900s. The evidence for the decline comes not from counts of hares but from game bag data. Declines have

Above: With its form in the shelter of a rock, this Mountain Hare is well hidden. Populations of Mountain Hares are stable.

Opposite: Habitat change, in particular agricultural intensification, threatens Brown Hares.

Above: Many Brown Hares are killed on the roads.

been recorded in the UK, Croatia, Denmark, Germany, Hungary, the Netherlands, Poland, Slovakia, Sweden and Switzerland. Several possible reasons for these declines have been suggested, including increased prevalence of disease, predation, shooting, urbanisation, expanding road and rail networks and adverse weather conditions. However the general Europe-wide intensification of agriculture seems to be the most likely overriding cause.

Agricultural intensification has led to: the increased use of agrochemicals, such as fertilisers and pesticides; the removal of hedges, other field boundaries and unfarmed 'fallow' areas of wild vegetation; bigger fields; more silage and less hay; more areas of monoculture; and increased mechanisation of farming (perhaps resulting in more leverets being killed by machines). Wild vegetation provides year-round shelter and food for hares. They like grass fields, and silage is usually cut many times each summer, while hay is cut at most twice. Each cut is dangerous for any leverets using the field. Intensification may result in fewer types of food being available for hares, and in 'gaps' in the seasonal food supply if only one or a couple of crops are grown and they all mature, are harvested or become unsuitable as hare food at the same time. Hares may not be able to find the patchy

Below: A Brown Hare in a village street. Urbanisation is leading to habitat loss for many wild species.

Above: Agricultural intensification has many facets. In combination they can have negative effects for lots of species of bird, insect, and mammal, including the Brown Hare.

clumps of vegetation they need to provide cover or shelter, and so may be more exposed to both the weather and predation.

There may seem to be a paradox: Brown Hares have declined due to agricultural intensification, but thrive in arable areas. In England at least, Brown Hares are most common in arable areas in the east, and remain rare in less intensively farmed areas in the west. But it's all relative: Brown Hares do best in areas of patchwork-like mixed farming, including big arable fields and some pasture, and with at least some unfarmed areas to provide permanent cover and food (wild plants) at lean times of year. Short vegetation (one study suggests about 8cm/3⅛in tall) is preferred for feeding, but areas with more cover are needed during the day, so mixed and less intensively farmed areas are best for Brown Hares.

Conservation and legal status

The IUCN recognises that the Brown Hare has declined throughout its range in Europe, but classes it as Least Concern. The Mountain Hare is also listed as Least Concern; the Irish Hare is listed in the Irish Red Data Book as an internationally important species, and was given an All-Ireland Species Action Plan in 2005. Both the Brown Hare and the Mountain Hare are listed as protected under Appendix III of the Bern Convention in Europe; the parties to this 'undertake to take all appropriate measures to ensure the conservation of the habitats of the wild flora and fauna species'. The Mountain Hare is listed in Annex V of the European Commission's Habitats Directive as a 'species of community interest whose taking in the wild and exploitation may be subject to management measures'. Member States must ensure that the exploitation and hunting of hares is compatible with maintaining them in a favourable conservation status.

Both Brown Hares and Mountain/Irish Hares were included in the UK Biodiversity Action Plan and were listed as UK priority species. Devolved plans in England, Scotland, Northern Ireland and Wales list the Mountain/Irish Hare for England, Scotland and Ireland, and the Brown Hare for England, Scotland and Wales.

Hunting and closed seasons

Hares are game animals that are hunted mainly by shooting. In many European countries, hares are protected by closed seasons when hunting is prohibited, severely limited or allowed only under licence. Most closed seasons are from January to September so that hares are protected during the breeding season, and hunting takes place from October until December, starting when populations reach their annual peak and leverets are no longer dependent on their mothers for milk.

In 2011, the Scottish Parliament introduced closed seasons from 1 February to 30 September for the Brown

Above: In most of Europe, Brown Hares are shot in autumn, at the end of the breeding season.

Hare, and from 1 March to 31 July for the Mountain Hare. In the Republic of Ireland there is a closed season for the Irish Hare from 1 March to 25 September; in Northern Ireland it is from 1 February to 11 August. In England and Wales, there is no closed season; the Hares Preservation Act 1892 prevents the sale of hares from 1 March to 31 July, but hares can be shot at any time. Traditionally, hare shoots take place in February, at the start of the breeding season, when many females are pregnant or suckling young. Shooting at this time is detrimental to the welfare of hares and can reduce local populations by 30–70 per cent.

As well as by shooting, hares have been hunted with dogs (packs of beagles hunting by scent, or greyhounds or lurchers hunting by vision during 'coursing'). Hare coursing was banned in Northern Ireland in 2010, but continues in the Republic of Ireland. In England and Wales, the Hunting Act 2004 controls the hunting of wild mammals with dogs and prohibits hare coursing. Hare hunting with dogs still takes place in France, but is illegal in most other European countries. In some countries, dogs are legally used to flush hares towards waiting hunters. Hunting with dogs has much less effect on numbers of hares than shooting does, but may cause disturbance and is considered detrimental for the welfare of hares.

In many individual game management areas, hunters themselves carry out counts or surveys of hares, assess the population and voluntarily restrict the numbers they kill. The existing closed seasons that protect hares when they are breeding must have conservation benefits as well as protecting the welfare of leverets, and may allow sustainable hunting. In England, without a closed season, many leverets are orphaned and die when their mothers are shot in February. A computer modeling study confirmed that hunting Brown Hares in February, as is traditional in England, results in many leverets being orphaned, and also in population declines. When the modeled population was protected during a closed season similar to that in Scotland, far fewer leverets were orphaned, and the population increased. It is clear that a closed season is needed in England and Wales.

How many hares?

For conservation planning, it's necessary to know how many individual animals exist and how numbers change over time. Estimating numbers on a large scale, for example in a country, is challenging, but numbers of Brown Hares in Britain have been estimated. An initial estimate published in 1992, based on game bag data and habitat preferences, was 1.3–2.0 million. Bristol University's national hare survey, conducted by volunteers walking through fields in daylight before the breeding season, produced an estimate of 817,520 for Britain, for the period of 1991–1993. This was extrapolated from 1,324 Brown Hares seen in 738 one-kilometre (2/3-mile) squares that were surveyed. The survey was repeated in 1997–1999; this produced an estimate of 732,000.

Right: In the light of a powerful torch, the eyeshine of this Brown Hare is conspicuous. Spotlight counting when hares are active at night is a good way to find out about numbers.

Above: Irish Hare populations have been monitored in recent years and seem to be thriving, but game bag data suggest that declines have occurred.

Since these attempts, the best estimates of hare abundance come from the British Trust for Ornithology (BTO). Volunteer surveyors conducting the BTO's annual Breeding Bird Survey in 2,600 one-kilometre (2/3-mile) squares record hares. The BTO data for 1995–2012 suggest that Brown Hare and Mountain Hare numbers have remained pretty stable.

The UK population of the Scottish Mountain Hare was estimated at 350,000 in 1995. A questionnaire survey conducted in 2006–2007 showed that Mountain Hares were present on about half the land surveyed, and that numbers taken by hunters were unlikely to cause declines in the population. About half the Mountain Hares killed were shot or snared for tick control; ticks on these hares are believed to affect Red Grouse by carrying the louping-ill virus, and grouse shoots are an important source of income.

Irish Hares are considered to have declined in the last 15–25 years – and evidence from game bag data suggests a longer decline, since the First World War. A daytime survey carried out in 1994 and 1995 in Northern Ireland showed that there were only 8,250 Irish Hares there. However, night-time surveys in 2002, 2004 and 2005 produced estimates of 7,000–25,200, 59,700–86,900 and 35,000–54,400 Irish Hares respectively. More recent surveys suggested that there were 47,000 (2006) and 113,000 (2007) Irish Hares in Northern Ireland, and 280,000 (2006) and 648,000 (2007) in all of Ireland. The picture is not clear, but the situation does not look terrible for the Irish Hare, and the regular surveys mean that changes over time can be monitored.

Hares released for restocking

Above: A hare being released – this one was released for research, rather than for restocking.

Many hares have been moved around by humans over the years, mainly for hunting. We know that Brown Hares can survive being moved because they exist outside their native range, for example in New Zealand, Argentina and Sweden, but many hares die during capture or shortly after release. As well as to start new populations, hares are moved around to boost existing populations by 'restocking'.

A study of hares released in Italy showed that, of 44 captive-bred hares, 38 died, most within 10 days of release. Many were eaten by Foxes or Beech Martens, and some died of coccidiosis or other diseases. A similar study in France showed that 32–53 per cent died within a week of release; most were killed or scavenged by Foxes, dogs, birds of prey or other predators, some were killed by traffic or farm machinery, and a few died due to disease. In Poland, 72 per cent of released hares died in the first year, most in the first month, and most were eaten by predators.

During the 1970s and 1980s, large numbers of wild and captive-bred hares from Bulgaria, Slovakia, Hungary and Poland were released in various parts of continental Europe. The lasting value of restocking for conservation

Above: This Brown Hare, photographed in Italy, has just left its form and is stretching its back legs, while remaining vigilant – one ear is pointing forward, one back.

is debatable, and it is certainly detrimental to the welfare of any unlucky wild hares that are caught for restocking purposes. Concerns about disease transmission, effects on the genetic structure of populations and welfare have resulted in a decline in hare restocking activities, yet they still continue. In parts of Italy, annual restocking is believed to be keeping the population of Brown Hares going, and restocking is still common in Greece, Poland, Switzerland and other countries.

Managing habitats to benefit hares

We know Irish Hares prefer farmed grassland, and Mountain Hares in Scotland prefer heather moorland, but little is known about how these habitats should be managed to benefit the hares using them. Increasing winter tourism (mainly skiing) in the Alps affects Mountain Hares there, and research is needed to monitor those populations. More attention has been given to managing habitats for Brown Hares.

Above: Mixed farmland, especially where it includes pastures, hay meadows, arable areas and some 'wild' zones, such as field margins, makes an excellent habitat for Brown Hares.

The key to habitat management for Brown Hares is to provide year-round cover and food. So, ideally pastoral beef and dairy farms should have some woodland, grassland and arable crops, and arable farms should have wheat, sugar beet and fallow land. Strips of uncultivated land in arable fields and windbreaks can benefit hares and increasing crop diversity also helps them. Hares seem to dislike the areas of short grass created by sheep grazing and so avoid heavily grazed sheep fields.

Hares may increase in number where gamekeepers remove their main predator, the Fox, but there is evidence from one study in Slovakia that managing habitats can increase hare numbers without killing predators. Here, treelines providing shelter, fallow land, game crops and more crop diversity were introduced on a 1,500-ha (3,710-acre) monoculture site over three years and spring hare density increased from 37 hares per sq km (96 per sq mile) to 67 hares per sq km (174 per sq mile). More recent research based on game bag data also suggests that habitat improvement can benefit hare populations.

The Brown Hare is not the only species believed to be declining mainly due to agricultural intensification, and many of the changes that benefit hares are likely to benefit other wild animals on farms farms too, including bats, farmland birds and insect pollinators. Pollination of

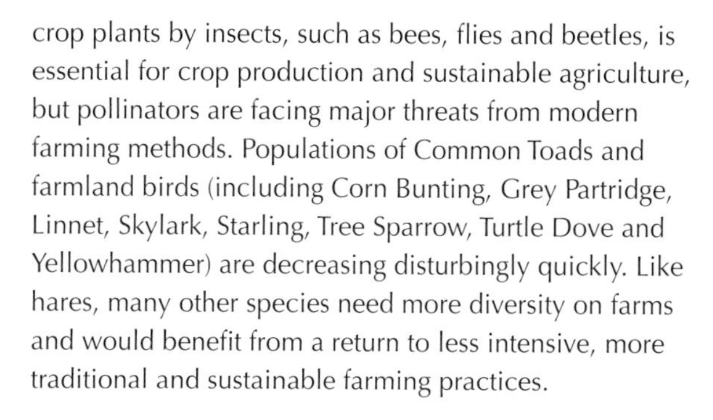

crop plants by insects, such as bees, flies and beetles, is essential for crop production and sustainable agriculture, but pollinators are facing major threats from modern farming methods. Populations of Common Toads and farmland birds (including Corn Bunting, Grey Partridge, Linnet, Skylark, Starling, Tree Sparrow, Turtle Dove and Yellowhammer) are decreasing disturbingly quickly. Like hares, many other species need more diversity on farms and would benefit from a return to less intensive, more traditional and sustainable farming practices.

Above left: Llike hares, Turtle Doves are believed to be declining mainly due to agricultural intensification.

Above: Pollinating insects are vital for crops, but they need diverse plants and habitats on farms to survive.

Below: Hares need predictable sources of water as well as food and shelter, especially in more arid parts of their range. This Brown Hare was photographed in Hungary.

Humans and Hares

Humans have interacted with hares for thousands of years, and it's clear that hares fascinate people. Historically an important source of food for many communities, wild hare is still enjoyed by many people today. Hares have also been used by humans in field sports, often to the hares' distress and detriment, and some people have, perhaps misguidedly, kept these beautiful wild animals as pets.

Hare products

Luckily for hares, they are hardly used in the fur industry, though hare skins can be used to make garments and their fur can be made into felt. Rabbit skins are used much more, and many Rabbits are bred only for their skins. Hairs from the ears, face and body of Brown Hares (and other species) are used to make fishing flies called 'hare's ears', which resemble the larvae of aquatic insects and are used to attract trout and other fish.

Most hares deliberately killed by humans are shot and eaten. Of the 32 hare species worldwide, the Brown Hare is one of the top three in terms of sheer numbers killed by humans. Around 70 million Brown Hares were exported from South America to Europe each year in the 1980s, and today over five million are killed each year in Europe. In Britain alone, hundreds of thousands of our hares are shot each year.

It is easy to understand the appeal of hare as a food source for humans. One hare can provide a substantial amount of meat, enough to feed perhaps eight people but, before you could cook and eat it, first you had to catch one. 'First catch your hare' is often claimed to be the first step in a recipe from Hannah Glasse's book *The Art of Cookery made Plain and Easy*, published in 1747, but in fact she didn't use that phrase. Mrs Glasse did warn her readers not to try to cook a hare with the skin

Opposite: An offering of a gazelle and a hare shown on a painting from the Tomb of Ounsou in West Thebes, Egypt, c. 1450 BC.

Below: Hunted Brown Hares in a game larder, ready to be used as food.

on: 'Take your Hare when it is cas'd [skinned], and make a pudding …'. Isabella Beeton, in her 1861 book *Mrs Beeton's Book of Household Management*, offered her readers tips on how to choose a young hare: 'The ears of a young hare easily tear, and it has a narrow cleft in the lip; whilst its claws are both smooth and sharp.'

Above: To make jugged hare, the meat is marinated in wine and cooked very slowly.

Hare meat is lean, dark and rich in flavour, so slow stewing is a good way to cook it, though young hares can also be roasted. A recipe for hare that has survived from Roman Britain involves stuffing the hare with pine kernels, almonds, spices, eggs and giblets, roasting it, and serving it with a sauce made from onions, dates, herbs and spices. Jugged hare, a traditional dish dating back to 17th century England (similar to the French *civet de lièvre* recipe) involves marinating hare in red wine with Juniper berries and cooking it slowly in a jug standing in a pan of water.

In Germany, *hasenpfeffer* is a traditional hare stew made from hare braised with onions, wine, vinegar, pepper and other spices. Similar recipes including paprika, sweet peppers and sometimes sour cream come from further east. Hare meat often features in game terrine or pâté, and the loin and back legs can be eaten fried, or made into a ragu and eaten with pasta (usually pappardelle). The modern trend is towards serving hare meat simply seasoned and grilled, to allow its unique flavour to be enjoyed.

It is of course desirable, though currently very difficult, to ensure that any hare we eat comes from a sustainably hunted population. Almost all countries in continental Europe, as well as Scotland, have a closed season for the hare, which at least ensures that they are not hunted at the start of the breeding season as they often are in England (see page 92–94). If a closed season is introduced in England, it will benefit the welfare of hares and go a long way towards ensuring that any hunting is sustainable, so that wild hares can thrive for many years to come.

The 'sport' of hare hunting

Hares hunted for sport, with the assistance of hawks, falcons or dogs, are sometimes eaten by the humans or the dogs. Poachers take hares illegally, without the permission of the landowner, probably mostly by lamping and shooting at night and with the help of lurchers (usually crosses between the sight hounds used for coursing and other working dogs). The effect of poaching on hare populations is unknown. Hares have also been taken by snaring – in nets placed across runs or gateways – by throwing hunting-sticks at their legs to make them tumble, by carefully aimed stones and by pitfall trapping.

There are two other ways in which dogs hunt hares. Hounds tracking hares by scent, mostly beagles or harriers, work in packs with humans following on foot or horseback. The dogs can take their time to find a hare and don't need to outrun it. Eventually the hare tires and they are able to catch and eat it.

Left: Hare hunting with a falcon, from an image published in Paris in 1864.

Below: Beagle packs were used to hunt hares in England until 2005.

Above: Greyhound racing, where the dogs follow a lure in the form of a mechanical 'hare', developed from hare coursing.

Above: Lady Florence Dixie, who called hare coursing 'torture'.

Hounds tracking hares by sight (coursing), often greyhounds or other sight hounds, have to be fast enough to outrun their quarry. Until it became illegal in 2002 in Scotland and in 2005 in England, coursing was organised as competitions between pairs of dogs and points were awarded for turning and killing the hare. It was practised as walked-up coursing, where a line of beaters would flush a hare from its form and two dogs were released to catch it; or as driven coursing, where hares were caught beforehand and later released for pairs of dogs to chase in front of spectators on a fenced course (as in the Waterloo Cup), with the hare given a head start of up to 100m (328ft). Coursing has an ancient history of more than 3,000 years, and if Brown Hares were introduced to Britain, they were probably introduced for the purpose of coursing. It is practised in many parts of the world, for example, in the USA, sight hounds are used to hunt jackrabbits (hares). Greyhound racing, using a fake mechanical 'hare' or lure, developed from coursing but is cruelty-free (for hares, at least).

Hunting hares with dogs is highly controversial and has been considered by some to be cruel and unnecessary for hundreds of years. In 1516, Thomas More wrote in *Utopia*: 'Thou shouldst rather be moved with pity to see a silly innocent hare murdered of a dog, the weak of the stronger, the fearful of the fierce, the innocent

Left: Hare coursing – two sight hounds compete to turn and kill the hare.

of the cruel and unmerciful. Therefore, all this exercise of hunting is a thing unworthy to be used of free men.' In 1892, British feminist and writer Lady Florence Dixie called hare coursing an 'aggravated form of torture'.

The League Against Cruel Sports was founded in 1924 and its aims included banning hare hunting and coursing. In England, the House of Commons voted to ban hare coursing in 1969 and 1975, but neither ban passed the House of Lords to become law. In 2002, the Scottish Parliament banned hare coursing in Scotland, and in 2004 the British Parliament passed the Hunting Act, which banned hare coursing and other forms of hunting with dogs from February 2005. The Waterloo Cup, which started in 1836, took place for the last time in 2005. It was the end of an era, but a victory for hare welfare.

Below: Coursing. A Brown Hare runs for its life, while a sight hound with a white collar competes against another with a red collar (not shown).

Hares as pets

Some people have kept hares as pets, though, like most wild animals, they are not particularly suited to it. Hares prefer to live in wide open spaces and so keeping hares confined in any way is unnatural; even in captive breeding centres, hares never really become tame. Sometimes people who come across leverets alone believe – rightly or wrongly – they have been orphaned and hand-rear them in order to rehabilitate and release them. This is controversial – people may consider it kind or interesting to 'save' an individual animal, but it is perhaps better for an orphaned leveret to remain in its natural habitat and part of the food chain as prey for a Fox.

The Belgian hare is a domestic rabbit breed that looks rather hare-like. If you love hares and wish you could have one as a pet, a Belgian hare is the best bet. These rabbits have been bred in captivity over many generations and are easier to care for and more suited to captivity than wild animals are.

Right: If they are released back into the wild, hand-reared hares like this one sadly have little chance of survival.

Above: A young Brown Hare being hand-fed on formula.

Below: The Belgian hare, a domestic rabbit breed, looks like a hare and makes a good pet.

Folklore and Art

Hares feature in folklore and art in many cultures, mostly as tricksters, witches, moon deities, or symbols of fertility, love, growth, dawn, rebirth and resurrection. Over the centuries, hares have somehow captured people's imagination and have been included in many phrases, fables, stories, paintings and sculptures.

Hares in language

People living in the countryside used many names to refer to the hare: 'puss' or 'wat' was common in southern England, 'maukin' or 'baud' in Scotland. A long list of names is given in the Middle English poem *The Names of the Hare*, including: old big-bum, hare-ling, frisky one, fast traveller, way-beater, white-spotted one, slink-away, nibbler, fellow in the dew, friendless one, cat of the wood, furze cat, stag of the stubble, lurker, skulker, low-creeper, sitter-still, one who turns to the hills, stag with the leathery horns, and get-up quickly.

Opposite: Painting of hares with Easter eggs, from Poland, around 1905.

Left: An illustration by Sir John Tenniel, from *Alice's Adventures in Wonderland* by Lewis Carroll, showing Alice with the March Hare and the Mad Hatter.

Above: This ancient Egyptian hieroglyph, showing a hare, means 'to be' or 'to exist'.

In ancient Egyptian, the hare hieroglyph means to be or to exist – perhaps an expression of the serenity of the hare. Hares pop up in several phrases and proverbs in the English language. For example, 'hare-brained' (foolish, impetuous or impulsive); 'to run with the hare and hunt with the hounds' (to support both sides of an argument or act hypocritically); 'to take hares with foxes', or 'to set the tortoise to catch the hare' (to try to do something impossible); 'to have two hares afoot' or 'to run after two hares' (to undertake too many things); 'to make a hare of' (to make ridiculous); 'there the hare went' (there the matter ended); 'first catch your hare' (don't make plans until you are prepared); 'as mad as a March hare' (crazy – first mentioned by Thomas More in 1529: 'As mad not as marche hare, but as a madde dogge'). To 'kill the hare' means to cut the last bit of corn in the field, and 'Hare and hounds' is a game in which a 'hare' lays the 'scent' (torn up paper) which the 'hounds' follow.

The Easter hare

In Anglo-Saxon paganism, as in other cultures, the fast-breeding hare is associated with fertility. The hare may also be a symbol of the spring goddess Eostre. Perhaps the similarity between a Lapwing's nest and a hare's form led to the belief that hares laid eggs. This may have resulted in the story of the Easter bunny bringing Easter eggs when, eventually, the pagan festival Ostara (on 21 March) became the Christian Easter.

In another explanation, the Easter bunny has a later origin. Christians in Germany saved their eggs during Lent, and hid them for their children to find at Easter. They invented stories of egg-bringers in the form of Foxes, storks, hares and other animals, but the hare eventually prevailed because it was featured in a popular book. The story spread to other parts of Europe, including the UK

Above: The Easter bunny bringing Easter eggs.

Left: A German Easter card from the 1960s. The text means 'Heartfelt Easter Greetings'.

and the British colonies, and evolved into the story of the Easter bunny. Over time, the eggs became chocolate and the hare became a cute bunny.

Hares in the moon and stars

In many cultures, the hare is closely associated with the moon, perhaps because it is like the hare: it waxes and wanes, and is associated with death, rebirth and madness (lunacy). In Eastern cultures people think they see a hare in the moon rather than a man. In China, the 'jade hare' in the moon is said to use a pestle and mortar to mix the elixir of immortality. Lepus, the hare, is a constellation south of Orion, the hunter; it is sometimes shown as being chased by Orion or his dogs.

Hares in art

Above: A postage stamp from Ireland, c. 1980, showing an Irish Hare.

Right: *Young Hare* by Albrecht Dürer, 1502.

The German artist Albrecht Dürer's lifelike painting of a hare, dated 1502, is perhaps the most famous hare painting, and is acknowledged as a masterpiece of observational art. Hans Hoffmann recreated it within his 1585 painting *A Hare in the Forest,* and Dürer's original has been reproduced many times, embossed in copper, wood and stone, represented in plastic and plaster, painted on eggs, copied as tattoos and printed on bags.

Right: *A Hare in the Forest* by Hans Hoffmann, 1585.

Hares feature in illustrated medieval manuscripts in relation to hunting, but also as tricksters – clever, but unconventional fictional characters that play tricks on others. In the *Decretals of Gregory IX*, commissioned in 1230 by Pope Gregory IX to document canonical law, hares are shown in an upside-down world, hunting and shooting dogs and humans.

Left: The Three Hares Roof Boss from the Chapter House at the Eglise Saints Pierre-et-Paul in Wissembourg, Alsace, France.

Below: The 'Window of Three Hares' in Paderborn Cathedral, Germany. The carving was created in the 16th century.

Pictures or sculptures showing three hares running in a circle are found in numerous locations, from the Far East to Devon in England. In this recurring motif the hares' ears touch in the centre of a circle so that, though there are only three ears, it appears that each hare has two. Around Dartmoor, these three hares are known as 'the Tinners' rabbits'. This enduring image is found mostly carved in stone inside churches, synagogues and other sacred sites, though it has been used in woodcarvings, stained glass windows, jewellery, coats of arms and tattoos, as well as in paintings and sculptures, and has been modified in lots of ways. The meaning of this ancient archetype, which traverses religions and cultures, has been lost over time.

Hares in fables

In the widely known Aesop's fable, a tortoise challenges a swift, cocky hare to a race. The tortoise wins due to his perseverance and because the hare, so confident that he will win, decides to take a nap during the race. The moral of Aesop's story is that endurance and tenacity can reap rewards.

Right: This 'tortoise and hare' sculpture in New York, USA, was designed by Michael Browne and erected in 1997.

In African folk tales the hare is presented as a trickster. Some of these stories were retold among African slaves in America, and evolved into the Br'er Rabbit stories of the southern USA. Brer Rabbit uses his cunning and wit to survive all manner of adventures. In the most famous story, *The Tar-Baby*, published in 1881, villanous Brer Fox makes a baby figure out of tar, and Brer Rabbit touches it and becomes stuck, as the fox intended. The more Brer Rabbit struggles, the more he sticks. Brer Fox is about to eat Brer Rabbit, when when Brer Rabbit tricks Brer Fox by suggestively begging Brer Fox not to fling him in the brier patch. Brer Fox sadistically throws the cunning rabbit into the brier, where the rabbit uses the thicket's branches to remove himself from the sticky tar baby and escape.

In a Russian folk tale, the old hunter Grandpa Mazai feels sorry for the hares that drown in the spring when the forest floods, so he goes out in his boat and saves them from precarious perches on logs, branches and

Left: Grandpa Mazai carrying hares to safety during a spring flood.

small islands. He releases them on higher ground, warning them that he, though he is saving their lives in spring, will hunt them in winter.

Jackalopes

The jackalope is a mythical creature in North American folklore: a hare with antlers (a sort of cross between a jackrabbit and an antelope). Jackalopes were popularised in the 1930s, when taxidermists in Wyoming sold jackalopes made by attaching deer antlers to stuffed jackrabbits. These bizarre horned rabbits are still made and sold today, along with postcards and other images of the creatures. They feature in poems, stories and video games, and there is a 'mockumentary' film about jackalopes. In Wyoming highway signs warn drivers to watch out for jackalopes and you can even buy a 'jackalope hunting license'!

The jackalope is not the only example of a hare with antlers. Horned hares were described (as '*Lepus cornutus*') in Europe and Asia over a period of many centuries and feature in medieval manuscripts as well as more recent scientific books. In Germany, the horned hare often has other strange body parts (wings and beaks), and is called the *Wolpertinger* (in Bavaria, southern Germany) or the *Rasselbock* (in Thuringia, central Germany).

This seemingly curious myth may have its origin in truth, as rabbits and hares can be infected with a papilloma virus that causes horn-like growths to form on their heads and bodies.

Below: A mythical jackalope, thousands of which are sold to tourists in the USA each year.

How to See Hares

For most people, getting a good view of any wild mammal is a rare and thrilling experience. Hares are elegant and beautiful, but also timid, secretive and elusive so trying to get a good view is worth the effort. A glimpse into the private life of the hare will open up a new and exciting world. If you aren't lucky enough to spot hares, the next best thing is discovering clear evidence of their existence. So keep your eyes down when you are in hare country – you might see a lovely soft form or a clear footprint.

Where to watch hares

If you are keen to see hares, it's best to head to a flat open area, preferably an arable field if you're looking for Brown Hares. For Mountain Hares in Scotland, your best chance will be in heather moorland, whereas for Irish Hares, any kind of agricultural land is a good option. Stick to roads or public footpaths, or seek permission from the landowner. European Rabbits and hares don't usually occupy the same areas, so if you see Rabbits or find Rabbit burrows in dry, sandy soil, you are unlikely to find hares nearby. Most farmers and other landowners know when they have hares on their land, so it's always worth asking them. And ask at nature reserves too – if hares are there, the reserve staff will know, and they may be able to tell you the best places to watch hares without disturbing them.

Opposite: A walker catches a glimpse of a Mountain Hare in the Cairngorms National Park, Scotland.

Below: Heather moorland (top right) is the best place to look for Scottish Mountain Hares. Brown Hares and Irish Hares can be seen in arable fields (bottom left and right) and in grassland.

Above: You might be surprised to learn that airports are often good places to see hares.

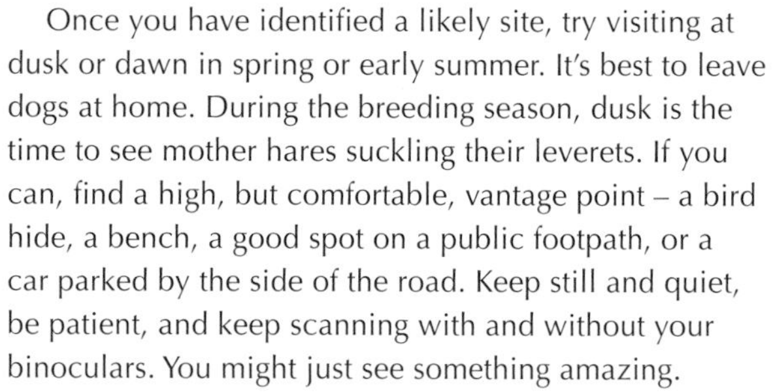

Hares are challenging to find, so allow plenty of time. In bright light, try scanning for hares in their forms by using binoculars. Hares may, at first, look like cow pats or molehills, but with practice you will be able to identify them. By walking around in the daytime, you may be able to flush an occasional hare from its form.

Once you have identified a likely site, try visiting at dusk or dawn in spring or early summer. It's best to leave dogs at home. During the breeding season, dusk is the time to see mother hares suckling their leverets. If you can, find a high, but comfortable, vantage point – a bird hide, a bench, a good spot on a public footpath, or a car parked by the side of the road. Keep still and quiet, be patient, and keep scanning with and without your binoculars. You might just see something amazing.

Below: A bird hide can help you get a good view of hares.

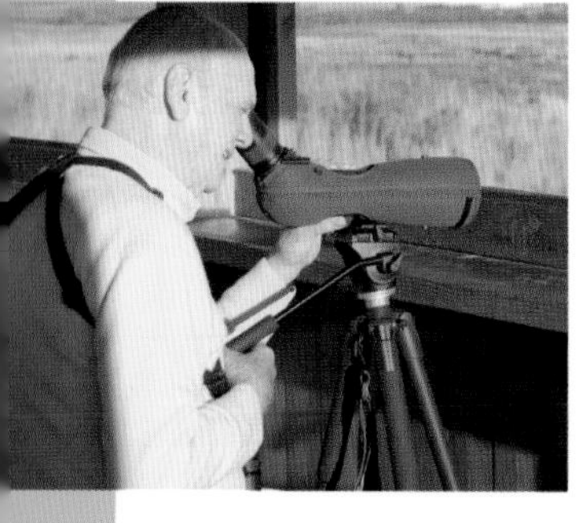

With permission from the landowner, you could also try lamping at night. First, make sure that people in nearby houses know what you are doing, or they may be alarmed or worry that you are poaching. You need a powerful torch – move it slowly and you may see hares

Above and below: If you see a hare during the day, you are likely to see more at the same place at night.

Above: An arable field before the crop gets too high is an attractive feeding spot for Brown Hares and therefore is also a good place to try and see them.

feeding in the beam of light. You are likely to see many more than you would during the day, as hares are likely to be feeding at night.

March to June is perhaps the best time of year to see hares – when the nights are short, so they are active in daylight, but the vegetation is not yet too high. If you are lucky enough to find a good place to watch hares, they will probably be there all year round. You might be able to see for yourself the annual cycle of mating, boxing, breeding and growing.

Tracks and signs

If you don't see hares, hares then finding proof that they have been around, in the form of field signs, is the next best thing. Brown Hares leave faint trails (paths) 10–20cm (4–7¾in) wide, but if you find one it is hard to be sure it is used by hares, and not Badgers, Foxes, deer or sheep. Tracks made by Mountain Hares are more obvious, and tend to go up and down hills, rather than along the contour; they are maintained by hares biting the vegetation.

Hares may leave footprints: impressions of the hind feet (6–15cm/2⅓–6in long, 3.5cm/1⅓in broad, with four toe prints visible) in front of those of the front feet

(4–5cm/1½–2in long, 3cm/1⅛in broad, five toes, though the inner toe or thumb may not be visible in prints). The hind feet are often side by side; the front feet are one behind the other. When hares are moving at speed, the hind feet pass the front feet in each stride. In good prints, the claws are visible. The prints of Scottish Mountain Hares and Irish Hares are very similar to, though slightly smaller than those of Brown Hares.

Rabbits' footprints are still smaller – the hind footprint is up to 4cm (1½in) long and 2.5cm (1in) broad. One way to tell them apart is by placing a matchbox (c. 5 x 3.5cm/ 2 x 1⅓in) over the hind footprint. If the matchbox is about the same width as the print, it is a hare's; if the print is about two-thirds the width of the matchbox, it is a Rabbit's. In hares, the hind feet are much longer than the front feet; in Rabbits, the hind feet are only about twice the length of the front feet, but often in prints the whole hind footprint is not visible. With luck, you might find prints made by a hare sitting on its haunches, and the full length of the hind feet will be visible; otherwise the whole foot shows only on very soft ground. The soles of hare and Rabbit feet are completely covered in hair, so their prints are unlike those of a dog or cat, in which clear indentations made by the pads of the feet can often be seen.

Below: Footprints in the snow are hard to identify, especially once the sun starts to melt them. These Mountain Hare footprints (**right**), photographed in the Peak District, UK, look very similar to Brown Hare footprints. It's unclear which species of hare made the footprints on the left.

Right: A snug form made by a hare in frosty grass.

If you are lucky, you might find a hare's form in vegetation or a ploughed field. Forms of Brown Hares may be dug up to 10cm (4in) into the ground, and are 10–20cm (4–7¾in) wide and about 25cm (9¾in) long. The deepest part of the form is the resting place for the hare's backside, and is often against a clump of grass, a stone or something else that provides shelter. Scottish Mountain Hares make well-defined forms, often in heather or rushes, located to ensure a good view.

Identification of droppings can be tricky: those of hares are 1.5–2cm (⅔–¾in) in diameter, slightly flattened brown or yellow-brown spheres, made of coarse plant fragments. Often they are found individually, though sometimes 10 or more are found in one place. The droppings of Rabbits are usually found in bigger piles, sometimes on raised hummocks, but they are smaller, about 1cm (⅓in) in diameter.

Below: Rabbit droppings (**left**) and Mountain Hare droppings (**right**) can be very hard to tell apart.

Above: A hare running off, showing the clear markings on its ears and tail.

Glossary

Caecotrophy Feeding on one's own droppings (caecotrophs) to maximise the extraction of nutrients from food.

Crepuscular Most active at dawn and dusk.

Dispersal One-way movement, usually undertaken by young male mammals, away from their place of birth.

DNA The genetic material contained within each cell of every living thing, and 'coding' for its characteristics.

Genus (plural genera) A group of very closely related species.

Herbaceous plants (or herbs) Plants that have no persistent woody stems. Herbaceous plants die after the growing season and new plants grow from seed (annuals), or die back and then re-grow from the root (perennials and biennials).

Hierarchy System in which members of a population are ranked by relative status or authority; the 'pecking order'.

Home range The area used by an individual animal or by a group of animals at a given time of the night, week or season. Home ranges are not defended and so may overlap.

Hybridisation Fertile sexual reproduction between two animals or plants of different species (or genera, subspecies, etc.). All the hare species in Europe can probably hybridise.

Invasive species A species, often non-native, that is harmful to its environment. Many species are not invasive in their native environment, but become invasive if they are moved to a new area.

IUCN International Union for Conservation of Nature (www.iucn.org). The IUCN produces the Red List of Threatened Species, which is used by scientists, conservationists and policy-makers.

Monoculture An area where only one crop is grown, so that over a large area there is little variation in available food or shelter for wild animals.

Nocturnal Most active at night.

Radio telemetry A method used to track individual animals by attaching a radio or Global Positioning System transmitter. Transmitters can provide other information, such as temperature.

Spotlight counting A method to assess numbers of nocturnal animals. A powerful torch is swung in a circle from fixed points, or moved along a line, and numbers of animals are counted, using the area of land lit up to work out their density.

Taxonomy The classification of living things into similar, related groups.

Territorial Having a territory, an area used exclusively by an individual animal or by a group of animals, and defended against intruders of the same species.

Acknowledgements

At Bloomsbury, I thank Jim Martin for suggesting me as a *Spotlight* author, and Julie Bailey, Jane Lawes and Katie Read for their editorial help. I thank Liz Drewitt at Nature Edit for her skilled and effective copy-editing.

My friends and hare experts Klaus Hackländer, Eric Marboutin, Silviu Petrovan and Philip Stott read and commented on my draft, making valuable suggestions and corrections. My dad, Richard Vaughan, inspired me to write.

Further Reading and Resources

BOOKS

The Nature Tracker's Handbook, N. Baker, 2013, Bloomsbury Publishing

Handbook of the Mammals of the World 6, Lagomorphs and Rodents, D. E. Wilson, T. E. Lacher Jr & R. A. Mittermeier, 2016, Lynx Edicions.

Handbuch der Säugetiere Europas 3/II, Hasentiere, J. Niethammer & F. Krapp, 2003, Aula.

Hare, S. Carnell, 2010, Reaktion Books.

Lagomorph Biology: Evolution, Ecology, and Conservation, P. C. Alves, N. Ferrand & K. Hackländer, 2008, Springer.

Mammals of the British Isles: Handbook, 4th edition, eds S. Harris & D. W. Yalden, 2008, Mammal Society.

Mein Name ist Hase, R. & V. Wirth, 2001, Staatliches Museum für Naturkunde Stuttgart.

The Hare, ed. A. E. T. Watson, 1896, Longmans, Green and Co.

The Hare Book, Hare Preservation Trust, 2015, Graffeg.

The Leaping Hare, G. E. Evans & D. Thomson, 1972, Faber & Faber.

The Way of the Hare, M. Taylor, 2017, Bloomsbury Publishing

ONLINE

The Mammal Society (www.mammal.org.uk)
The Mammal Society is a charity advocating science-led mammal conservation. The Society leads efforts to collect and share information on mammals, to encourage research to learn more about their ecology and distribution and to contribute to their conservation.

The Hare Preservation Trust (www.hare-preservation-trust.co.uk)
The Hare Preservation Trust aims to raise the profile of the Brown Hare and Mountain Hare and campaigns for legal protection and conservation.

The Irish Hare (www.irishhare.org)
This website provides information on hares in Ireland, campaigns for their preservation, and background on hares in Irish folklore.

Image Credits

Bloomsbury Publishing would like to thank the following for providing photographs and for permission to reproduce copyright material.

While every effort has been made to trace and acknowledge all copyright holders, we would like to apologise for any errors or omissions and invite readers to inform us so that corrections can be made in any future editions of the book.

Key t= top; l= left; r= right; tl= top left; tc= top centre; tr=top right; cl= centre left; c= centre; cr= centre right; b= bottom; bl= bottom left; bc= bottom centre; br= bottom right

AL = Alamy; FL = FLPA; G = Getty Images; NPL = Nature Picture Library; RS = RSPB-images.com; SH = Shutterstock

Front cover t Chris Upson/G; **b** Andy Rouse/G; **spine** David Tipling/G; **back cover t** David Tipling/G; **b** Les Stocker/G; **1** BMJ/SH; **3** CaryfukEOS/SH; **4** Ana Gram/SH; **5** Anton Gvozdikov/SH; **6t** The Natural History Museum/AL; **6b** The Natural History Museum/AL; **7l** Matthijs Wetterauw/SH; **7r** Peter Wey/SH; **8t** Neil Burton/AL; **8b** Movie Poster Image Art/Contributor/G; **9t** blickwinkel/AL; **9b** Andy Rouse/NPL; **11** Mirko Graul/SH; **12** Neil Bowman/FL; **13** Peter Wey/SH; **14** Ross Kinnaird/Staff/G; **15** Mark Medcalf/SH; **16** Mark Medcalf/SH; 17t Mark Medcalf/SH 17b Terry Andrewartha/FL; **18tl** Laurent Geslin/NPL; **18tr** Natursports/SH; **18b** Andrew Parkinson/RS; **19l** Michael Westerop/SH; **19cl** Mark Hamblin/2020VISION; **19cr** Ross Kinnaird/Staff/G **19r** kwhw/SH; **20** Emi/SH; **21** John Cancalosi/NPL; **23** Pyshnyy Maxim Vjacheslavovich/SH; **24t** Tom Ennis; **24c** Ivaschenko Roman/SH; **24b** Knumina Studios/SH; **26** allanw/SH; **27t** Inge Jansen/SH; **27b** Andy Rouse/NPL; **28l** Digoarpi/SH; **28r** nomad-photo.eu/SH; **29t** Soru Epotok/SH; **29b** Bildagentur Zoonar GmbH/SH; **30** Adwo/SH; **31** Georgios Kollidas/SH; **32** David Kjaer/NPL; **34t** Stritchy/SH; **34b** A. Cambone, R. Isotti – Homo ambiens; **35t** Wildscotphotos/AL; **35b** José Mª F. D. Formentí/www.formentinatura.com; **36l** Tom Reichner/SH; **36r** Paul Sawer/FL; **37t** Verónica Farías UNAM; **37b** Photo Researchers/FL; **38** David Welling/NPL; **39** Platsiee/SH; **40t** Endangered Wildlife Trust's Drylands Conservation Programme; **40b** Leibniz Institute for Zoo and Wildlife Research/WWF Vietnam-CarBi/Quang Nam Saola Nature Reserve; **41t** Randimal/SH; **41b** Roland Seitre/NPL; **42** Ian Maton/SH; **43** Weidong Li; **44** Dennis Jacobsen/SH; **45** Mogens Trolle/SH; **47t** Losonsky/SH; **47b** PATRICK PLEUL/Stringer/G; **48tl** BMJ/SH; **48tr** Abi Warner/SH; **48b** Martin Fowler/SH; **49** Richard Packwood/RS; **50** nomad-photo.eu/SH; **51** Red Squirrel/SH; **52t** David Tipling/NPL **52b** Klaus Hackländer; **53** Mark Caunt/SH; **54** Barcroft/Contributor/G; **55** Gail Johnson/SH; **56** Giedriius/SH; **57** Phil McLean/FL; **58t** Simon Litten/FL; **58b** Christophe Rolland/SH; **59** bikeriderlondon/SH; **60** Barcroft/Contributor/G; **61** Panos Karos/SH; **62** Laurie Campbell/SH; **63t** Terry Andrewartha/NPL; **63b** Chris Knights/RS; **64** Pavel Krasensky/SH; **65** Arterra/Contributor/G; **66** Sergei Ivanov1/SH; **67t** Liga Gabrane/SH; **67b** Andy Rouse/NPL; **69** Andy Rouse/NPL; **70t** Jim Brandenburg/Minden Pictures/FL; **70b** Klaus Hackländer; **71** Martin Hesko/SH; **73** Paul Sawer/FL; **74** Mark Sisson/RS; **75** bikeriderlondon/SH; **76** Neil Burton/SH; **77** Kevin Sawford/RS; **78t** Martin Fowler/SH; **78b** images72/SH; **79** Andrew Mason/FL; **80t** Ben Queenborough/G; **80b** Paul Sawer/FL; **81t** Andrew Astbury/SH; **81c** Jamie Hall/SH; **81b** bikeriderlondon/SH; **83t** defotoberg/SH; **83b** Colin Pickett/AL; **84** my nordic/SH; **85tl** Piotr Krzeslak/SH; **85tr** Mark Bridger/SH; **85bl** Angelo Gandolfi/NPL; **85br** Ondrej Prosicky/SH; **86** Red Squirrel/SH; **87** Volodymyr Burdiak/SH; **88** Jamie Hall/SH; **89** Rolf Giger **90t** Regien Paassen/SH; **90b** Mirko Graul/SH; **91** Frank Rumpenhorst/Staff/G; **93l** Konjushenko Vladimir/SH; **93r** Corepics VOF/SH; **94** dpa picture alliance/AL; **95** Carl Morrow/AL; **96** Klaus Hackländer; **97** DEA/G.CAPPELLI/Contributor/G; **98** David Hughes/SH; **99tl** xpixel/SH; **99tr** Happy Owl/SH; **99b** MZPHOTO.CZ/SH; **100** De Agostini Picture Library/Contributor/G; **101** Fotosenmeer/SH; **102** Simon Reddy/AL; **103l** Marzolino/SH; **103r** Paul Wishart; **104t** Bettmann/Contributor/G; **104b** Artokoloro Quint Lox Limited/AL; **105t** Universal History Archive/Contributor/G; **105b** Terry Andrewartha/NPL; **106** Sean Hunter/FL; **107t** Sean Hunter/FL; **107b** Francis Apesteguy/Contributor/G; **108** Neirfy/SH; **109** Morphart Creation/SH; **110** The Art Archive/AL; **111l** Kohl-Illustration/AL; **111r** Christos Georghiou/SH; **112tl** Boris15/SH; **112tr** Fine Art/Contributor/G; **112b** Fine Art/Contributor/G; **113t** Chris Chapman, The Three Hares Project; **113b** LOOK Die Bildagentur der Fotografen GmbH/AL; **114** Joseph Sohm/SH; **115t** SPUTNIK/AL; **115b** Diana Haronis/G; **116** Mark Hamblin/2020VISION/NPL; **117t** DrimaFilm/SH; **117bl** Emjay Smith/SH; **117br** Tim Mountford – JMS Group/SH; **118t** Andy Rouse/NPL; **118b** Marisina/SH; **119t** Frederic Desmette/G; **119b** David Tipling/RS; **120** Chrispo/SH; **121l** Ullamaija Hallinen/Folio Images/G; **121r** Nature Picture Library/G; **122t** Paul Sawer/FL; **122bl** Emily Veinglory/SH; **122br** Bob Gibbons/FL; **123** wim claes/SH

Index